Springer Theses

Recognizing Outstanding Ph.D. Research

Aims and Scope

The series "Springer Theses" brings together a selection of the very best Ph.D. theses from around the world and across the physical sciences. Nominated and endorsed by two recognized specialists, each published volume has been selected for its scientific excellence and the high impact of its contents for the pertinent field of research. For greater accessibility to non-specialists, the published versions include an extended introduction, as well as a foreword by the student's supervisor explaining the special relevance of the work for the field. As a whole, the series will provide a valuable resource both for newcomers to the research fields described, and for other scientists seeking detailed background information on special questions. Finally, it provides an accredited documentation of the valuable contributions made by today's younger generation of scientists.

Theses are accepted into the series by invited nomination only and must fulfill all of the following criteria

- They must be written in good English.
- The topic should fall within the confines of Chemistry, Physics, Earth Sciences, Engineering and related interdisciplinary fields such as Materials, Nanoscience, Chemical Engineering, Complex Systems and Biophysics.
- The work reported in the thesis must represent a significant scientific advance.
- If the thesis includes previously published material, permission to reproduce this must be gained from the respective copyright holder.
- They must have been examined and passed during the 12 months prior to nomination.
- Each thesis should include a foreword by the supervisor outlining the significance of its content.
- The theses should have a clearly defined structure including an introduction accessible to scientists not expert in that particular field.

More information about this series at http://www.springer.com/series/8790

Yifan Zeng

Research on Risk Evaluation Methods of Groundwater Bursting from Aquifers Underlying Coal Seams and Applications to Coalfields of North China

Doctoral Thesis accepted by
the China University of Mining and Technology, Beijing,
China

Author
Dr. Yifan Zeng
College of Geoscience
and Surveying Engineering
China University of Mining
and Technology
Beijing
China

Supervisor
Prof. Qiang Wu
College of Geoscience
and Surveying Engineering
China University of Mining
and Technology
Beijing
China

ISSN 2190-5053 ISSN 2190-5061 (electronic)
Springer Theses
ISBN 978-3-319-79028-2 ISBN 978-3-319-79029-9 (eBook)
https://doi.org/10.1007/978-3-319-79029-9

Library of Congress Control Number: 2018937325

Printed on acid-free paper

This Springer imprint is published by the registered company Springer International Publishing AG part of Springer Nature
The registered company address is: Gewerbestrasse 11, 6330 Cham, Switzerland

Supervisor's Foreword

Mine water inrush events often occur during coal mine construction and production; they account for a large proportion of the nation's coal mine disasters and accidents in China. Between 2005 and 2014, 513 water inrush incidents have occurred with a total loss of 2753 lives. As mining depths and mining intensity continue to increase, the hydrogeological conditions encountered are becoming more complex. One challenge is to prevent or predict water inrushes from the aquifers that underlie many of the coal seams. Because water inrush from the underlying aquifers is a nonlinear dynamic process, its occurrence is controlled by multiple factors and involves complex mechanisms. Dynamic nonlinear processes are not readily amenable to mathematical equations. The water inrush coefficient, introduced in the 1960s, has been widely used by coal mine hydrogeologists because it had the advantages of being a simple physical concept, convenient to calculate, and easy to use. It has been modified several times to better reflect actual water inrush conditions and has played a positive role in resolving the dangerous problem of water inrush from underlying aquifers in China. However, the water inrush coefficient method only considers two factors: the potentiometric pressure of the underlying confined aquifer and the thickness of the aquitard that functions as a water barrier between the coal seam and the underlying aquifer. Other factors also govern water inrush from underlying aquifers. In addition, the water inrush coefficient threshold is empirical and typically determined using reported water inrush incident statistics. Because geological and hydrogeological conditions can vary significantly in different areas, considerable deviations can exist between results of water inrush assessments and reality.

In this thesis, Dr. Yifan Zeng developed an information fusion model with information fusion theory, geographic information system technology, and modern mathematical methods to evaluate the risks of groundwater inrushes from aquifers underlying coal seams. In the new model, the water inrush vulnerability index was calculated with a variable weights theory. This method overcomes the defect of the traditional vulnerability index method in which weights of the water inrush main factors are constant in the evaluation process. The innovative model was applied to two coal mines in China. The results proved better than the traditional vulnerability index method. The work reported in this thesis represents a significant advance in addressing the global issue of mine water control and management.

I congratulate Yifan for this excellent work. His dissertation is one of the best in China University of Mining and Technology, Beijing, because of the volume of reliable data, defensible scientific analysis, and world significance of the research results.

Beijing, China
December 2017

Prof. Dr. Qiang Wu

Parts of this thesis have been published in the following journal articles:

Y Zeng, S Liu, W Zhang (2016) Application of artificial neural network technology to predicting small faults and folds in coal seams, China. Sustain. Water Resour. Manag. 2(2):175–181

Y Zeng, Q Wu, S Liu (2016) Vulnerability assessment of water bursting from Ordovician limestone into coal mines of China. Environ. Earth Sci. 75(22):1431. doi:10.1007/s12665-016-6239-4

Y Zeng, Q Wu, S Liu (2017) Evaluation of a coal seam roof water inrush: case study in the Wangjialing Coal Mine, China. Mine Water Environ. doi:10.1007/s10230-017-0459-z

This research was supported by the following projects

China National Scientific and Technical Support Program (2016YFC0801801).

Establishing Planning of National Engineering Research Center of Coal Mine Water Hazard Control in 2014 (2014FU125Q06).

China National Natural Science Foundation (41572222, 41702261, 41702270, 51774136).

China Postdoctoral Science Foundation (2016M601172).

The Key Laboratory of Karst Environment and Geohazard, Ministry of Land and Resources (KST2017K02).

The Hebei State Key Laboratory of Mine Disaster Prevention (KJZH2016K01).

Wuhan Yellow Crane Talents Program.

Contents

Chapter 1
Introduction

1.1 Research Background and Purposes

1.1.1 Present Situation of Water Hazards in Coal Mines of China

The coal resource is abundant in China. By the end of 1988, the proved reserves were nearly 400 billion tons according to the statistical results of the "1989 World Energy Conference", ranking second in the world [1] and 30 billion tons less than the statistics published by the United States. With the rapid development of Chinese economy in recent years, the coal production has been ranked No. 1 in the world. For example, from 2010 to 2015, the average annual output reached 3.62 billion tons, accounting for half of the world's annual production [2]. However, the hydrogeological conditions in the coalfields of China are complex, and the types of water hazards are diverse. Because of the number of coal mines, the amount of coal reserves and the severity threatened by water bursting, preventing and controlling water have been a challenge. The water hazards mainly occur in the coal mines in south China and north China. The north China coal ore districts are mainly threatened with Carboniferous and Permian karst fissure water hazards while the coal fields in south China are mainly threatened with late Permian karst water. In the northwest coal-producing area, the Jurassic fissure water hazard is the main factor. In the northeastern part of China, there are not only sandstone fissure water hazards of the coal seam floor but also the hazards caused by the Quaternary pore water [3]. In addition, due to the long history of coal mining in China, old kilns and goafs are present in the shallow areas of many mine and the surrounding areas. Almost every ore district in south China and north China is faced with the threat of pooled water in abandoned kilns and goafs. In recent years, water disasters caused by old kilns have shown an upward trend.

Coal mine water inrushes and flooding accidents occurred frequently in China. Over the past decade, the number of heavy, large water inrush accidents was 513, which is unprecedent. A total of 2753 people was injured or lost their lives because of these water hazards (Table 1.1). In 21 days from April 6, 2012 to April 26, 2012, five large, major coal mine flooding accidents occurred with 49 fatalities [4, 5]. 12 people died in the water inrush accident in Jilin Fengxing Coal Mine on April 6,

Y. Zeng, *Research on Risk Evaluation Methods of Groundwater Bursting from Aquifers Underlying Coal Seams and Applications to Coalfields of North China*, Springer Theses, https://doi.org/10.1007/978-3-319-79029-9_1

Table 1.1 Select water disasters in coal mine from 2005 to 2014

Year	Number of water disasters	Number of fatalities
2005	104	593
2006	38	267
2007	63	255
2008	49	263
2009	21	125
2010	18	167
2011	22	163
2012	13	107
2013	14	245
2014	13	217
Total	513	2753

2012. On April 10, 2012, four people were killed because of the water inrush in Kongzhuang Mine. On April 13, 2012, 11 people lost their lives in the accident of water inrush in Changzhifulianying Mine [5–7].

In addition, the water inrush (burst) and inundation accidents often resulted in huge economic losses and casualties. For example, in Henan Jiaozuo coalfield, there were nine significant water hazard accidents with bursting water quantity up to 60 m^3/min, resulting in a total mine discharge of 540 m^3/min. They not only caused serious over-exploitation of groundwater and waste of water resources, but also resulted in the high drainage cost in per ton coal, accounting for approximately 20% of raw coal prices. The annual drainage costs of the mine were 20–25 million Yuan in Renminbi (RMB) [8]. In 1984, two mines were flooded, and one mine was forced to stop production in the Fangezhuang Mine water inrush accident, resulting in direct economic losses up to 0.5 billion RMB. In 2005, the mine flooding accidents in Meizhou Daxing Coal Mine and Handan Niuer Zhuang Mine caused economic losses up to 47.25 million and 382 million RMB, respectively [9]. In 2005 alone, there were 109 coal mine accidents in China, killing 605 people, including 13 large-scale water hazard accidents with 360 deaths. On August 7, 2005, Meizhou Daxing coal mine flooding accident resulted in six people killed and 117 people missing [10].

1.1.2 Current Situation of Water Hazards from Coal Floor in Coalfields of North China

North China Carboniferous Permian coalfields are the most important coal-bearing regions and coal production bases in China. They are distributed in more than 70 ore districts in 12 provinces such as Shanxi, Mongolia, Hebei, Shandong and Henan. The coal reserves and production account for approximately 68 and 65% nationwide, respectively [11, 12]. Maintaining the stability of the coal production scale is of great importance to China's economic and social development. However, the hydrogeological conditions of the north China coalfields are extremely complex and vary. The water hazard accident has become the second killer of the coal mines only after the gas accidents. The water hazards seriously threaten the safe production of the coal mines.

Water discharge and bursting into mines occurred frequently in north China coalfields. When the water yields were large, they seriously affected the normal production of coal mines. In addition, the coal mines had to pay enormous cost of drainage, which not only increased the cost in per ton of coal but also caused a great waste of the valuable groundwater resources. The coal seams in the middle and deep of the north China coalfields were threatened by the strong aquifers at their bottom. Tens of billions of tons of safe pillars could not be recovered [13, 14].

High stress areas were formed, and the coal seam pressure became higher and higher in north China coalfields as the mine was gradually developed deeper and deeper. The underlying karst aquifer of the middle-deep lower coal group was supper thick because of deposition. Due to the multi-period crustal movement and post-tectonic transformation, the karst aquifer was seriously damaged, resulting in a high degree of karst development. The aquifer was not only rich in water but also highly permeable. The effective thickness of confining bed between the karst aquifer and the middle-deep lower coal group got smaller and smaller due to the disturbance of the mine engineering. Pressure on the coal seam floor was also increased as coal mining approached to the deep constantly. Under the combined influence of several negative factors, like high pressure, high water abundance and thinner thickness of confining bed, the coal exploitation became more and more difficult. The risk of coal seam floor water inrush significantly increased with mining space and degree of mechanization improved [15–17]. Therefore, it is of great practical significance to study the targeted prevention and control measures to relieve the threat of water hazard and to safely and efficiently exploit the lower group coal [17, 18].

1.2 Research Status of the Subject

1.2.1 Current Situation of Research in the World

The development of coal industry provides a basis for the study of coal seam floor water inrush. The history of large-scale development of the coal industry in the former Soviet Union, Hungary, Germany and other countries has been nearly a hundred years. In the coal mining practices, both experience and technologies in the management of floor water hazard have been accumulated and several theoretical doctrines have been brought up.

In 1944, Wegger Frens, a Hungarian and the originator of studying the floor water hazard, based on static mechanics, gave the definition and calculation formula of the relative aquifer [19, 20]. Using water pressure and the thickness of floor aquifuge to quantitatively assess the possibility of water inrush opened the prologue to the study on water hazard from coal floor, and the associated floor water inrush assessment methods had been applied in China until now. In the same period, Slalesh Lev, based on the idea of material mechanics, proposed a simplified calculation model of floor water inrush, that is, the floor aquifuges were idealized as fixed beams on both sides and the water pressure were assumed to be the stress acting on beams [21], and then a calculation formula of safe hydrodynamic pressure for safe mining was deduced.

In the 1960s, the Hungarian had officially written the technical method of using relative thickness of the aquifers to assess floor water inrush into national norms, named "Mining Safety Regulations". Different statements of calculation were illustrated according to different water and hydroelectric geological conditions. Over this period, various kinds of water hazard from coal floor that were encountered during the development of the coal industry promoted coal workers to pay attention to and study the floor water problems based on the static mechanics theory.

From 1970s to the 1980s, the study of coal floor water hazard began to introduce the research results of rock mechanics. In the study of the failure mode of the waterproof rock group, Santos concluded the strength criterion of Hoek-Brown rock mass. The energy release point was used to quantitatively determine the compressive stress of the waterproof rock group [22, 23]. Sammarco found that early warning of water bursting can be achieved by monitoring precursor information of water inrush such as rapid changes in water levels [24, 25]. V. Mironenk firstly proposed that water-resisting layer would be destroyed by mining disturbance when he studied the floor water hazard [26, 27]. He believed that the floor water gushing-out was the result of water resisting floor to be broken under the action of mining disturbance and water pressure.

After the 1980s, the seepage theory of groundwater in the rock fracture has made great progress since the discovery of the cubic law of the permeability coefficient of the plane fracture network. Erichsen constructed a coupled model of groundwater seepage-geostress interaction [28, 29]. Elsworth built a flow-fluid-solid coupling model for calculating the bias current of groundwater in rock fissures [30, 31]. All the above results were absorbed into the floor water inrush research, improving the research on floor water permeability mechanism and water hazard prediction method.

1.2.2 Current Situation of Research in China

Before 1949, the hydrogeological work in the mining areas was very weak. Although coal production was low and mining depth was shallow, mine flooding accidents happened quite often. In 1935, the catastrophic water inrush accident in Zibo North Mine resulted in 350 deaths, and the entire mine was flooded and scrapped. After 1949, the coal industry has been rapidly developed. The scale, scope and depth in coal mining continuously extended with the progress of mining technology. Meantime, the encountered problem of floor water hazard became increasingly prominent, and the hydrogeological work of the mine became an indispensable part in the coal mine

Table 1.2 Water inrush coefficient (T_S) and its evolution history

Time	1960s	1970s	1980s	2005	2009
Formula	$T_s = \frac{P}{M}$	$T_s = \frac{P}{M-C_p}$	$T_s = \frac{P}{\sum_{i=1}^{n} M_i m_i - C_p}$	$T_s = \frac{P_c}{M \cdot \sum_{i=1}^{n} \xi_i M_i - M_i - M_m}$	$T_s = \frac{P}{M}$
Symbol description	P: water pressure, MPa; M: thickness of water resisting layer, m	C_P: depth of coal seam destroyed by mine pressure, m	M_i: thickness of the i-th water resisting layer, m m_i: conversion coefficient of equivalent thickness of the i-th water resisting layer	P_c: residual water pressure in intrusion point on water head intrusion belt, MPa; ξ_i : equivalent anti-hydropressure intensity factor of different stratum; M_l: Destruction zone of seam floor, m; M_m: thickness of water head original intrusion belt, m	M: thickness of effective water resisting layer of coal seam floor, m
Rationales for modification	Proposed on water prevention cons odium at Jiaozuo	After the impact of mining was considered	After stratified aquifuges and mechanical property were considered	Floor failure, separation group of confining beds, confined water intrusion considered	Revised in new regulations

development. Study on the coal seam floor water inrush mechanism, prediction and evaluation also gradually grew and developed. Many research results were accumulated, as summarized in the following sections.

1.2.2.1 Empirical Formula

(1) Water inrush coefficient method

In 1964, water inrush coefficient was put forward, in the meeting organized by the Ministry of Coal Industry to tackle key problems of water inrush in Jiaozuo mine, according to the statistical data of water pressure on the bottom of aquifuges and thickness of aquifuges from these main water abundant ore district in north China coalfield, based on the concept of relative aquifuges in Hungary. The physics conceptual model of hydrogeology of water inrush coefficient method is simple and easy to use. Although it has been modified many times (Table 1.2), it is still in use today [32–37].

(2) "Down Three Zone" theory

In the early 1980s, the Jingxing Mining Bureau and the Shandong Mining Institute divided the water resisting floor into upper damage zone and lower affected zone according to the degree of damage of water resisting floor disturbed by mining.

In 1988, Shandong Institute of Mining, through years of research, summed up the "Down Three Zone" theory [38–41]. The water resisting layers from top to bottom were divided into zone destroyed by mining disturbance, complete rock zone and confined water guiding zone after the consideration of the damage effect of the mining disturbance to the coal seam floor. It was concluded that once the complete rock zone was destroyed, coal seam floor water inrush would occur.

(3) "Progressive-Intrusion" theory

Wang Jingming's "Progressive-Intrusion" theory was based on fluid-solid coupling model, and he believed that the groundwater would pour into fissures within water resisting layers under the effect of water pressure. When the coal mine was exploited, fissures would be further developed, and groundwater would be further upward under the combined action of mining disturbance and water pressure. Groundwater would burst into the mines along the fissure channel once these fissures connected to the bottom of the zone destroyed by mine pressure [42, 43].

1.2.2.2 Water Inrush Theory Based on Rock Mass Mechanics

(1) Plate theory

In 1977, Liu Tianquan thought that under the influence of mining disturbance, water-proof rock group could be regarded as a simple model composed of water-resisting layer and fissure zone. The water-resisting layer was assumed as a force sheet with four-sided fixed made of elastic material [44–46]. A method to calculate the maximum bearable hydrostatic pressure was deduced. In this theory, the concept model of hydrogeology was simple, and its calculation method was convenient, but the biggest drawback was that the ratio of thickness to width of floor water-resisting zone was difficult to meet the prerequisite of thin plate theory.

(2) "Zero position damage" or "In situ fracture" theory

In 1992, Wang Zuoyu, Liu Hongquan divided the coal road into three sections, namely, stress compression section which had not been mined in the front part (I section), surrounding rock expansion section of the middle part (II section) and caving stability zone in the back part in accordance with the impact of coal mining on the coal road. They suggested that under the combined action of water pressure and ground pressure, "in situ fracture" was produced in the floor rock mass in the lower part of I and II section [47–49]. Stress release from floor rock due to coal mining caused "Zero position damage" of floor rock mass in the lower part of the II and III sections [50].

(3) "Strong seepage channel" theory

In the early 1990s, Xu Xuehan believed that conducting-water structure was the main factor of the occurrence of water hazard. The water inrush mechanism of original and secondary conducting-water structure, the secondary one caused by the disturbance of mining engineering, was respectively analyzed. This theory emphasized the influence of geological structure on floor water inrush [51, 52].

(4) "Stress relationship between rock and water" theory

In the late 1990s, Wang Chengxu simplified the floor water inrush model to the interaction result of rock, water and stress. There were water-flowing fractured zones developed in confining beds, and they were further expanded under the action of water, rock and mining engineering [53–55]. When the water pressure was greater than or equal to the minimum principal stress confining bed group, the water inrush would happen.

(5) Key strata (KS) theory

Qian Minggao and Li Liangjie considered a set of complete water-resistant rock with the biggest stress tolerance located on water filling aquifer as "Key Strata". This set of complete water-resistant rock was supposed as a four-side fixed thin plate, and once this thin plate was destroyed under stress, water inrush would occur. The maximum rupture span of floor KS was obtained according to the theory of sheet strength [56–62].

1.2.2.3 Pan Decision Analysis Theory

The definition of pan-decision analysis methods varies based on the use of mathematical technology model. The central idea of this theory is using some soft science method model to solve the correlative problem of every influencing factor based on system theory, so water inrush forecast is taken as a response to the impact of factors. This method is widely used in the prediction of floor water inrush. For example, Zheng Shishu (1994) used geographic information system (GIS) technology to establish a water inrush prediction model, taking into account water pressure and the thickness of aquifer and other factors. Shao Aijun (2001) used the principle of mutation to forecast the floor water hazard in the Xingtai ore district. Li Fuping (1997) and Jin Dewu (1998) applied ANN technology to couple all the influencing factors and discussed methods for predicting water inrush from coal mine floor. However, there are errors with the evaluation results to the actual situation of the mines because the selected influencing factors for all these above-mentioned evaluation methods are not comprehensive enough and the relative importance of the influencing factors on the water inrush are basically determined by personal experience [63–73].

1.2.2.4 Vulnerable Index Method

In the 1990s, Wu Qiang proposed the vulnerability index of coal seam floor water hazard evaluation [74, 75]. Before the vulnerability index method was put forward, the most commonly used method for water inrush evaluation was "water inrush coefficient method" [76]. The water burst coefficient method played an important role in guiding the safe exploitation of coal in China. However, its physics conceptual model of hydrogeology was too simple to truly reflect the complex mechanism and evolution of floor water inrush under the action of various influencing factors [77].

Vulnerability index method is prediction and evaluation method of water inrush from coal seam floor by coupling the information fusion method, which can determine the weight coefficients of multi main controlling factors for assessing floor water inrush, with the GIS with strong spatial information analysis and processing functions [78]. It can be used in predicting the risk of water inrush, under the combined effect of ground pressure and water pressure, from coal seam floor which are made by multi-lithologic and multi-rock formation and under the influence of different types of tectonic destruction. This method is still constantly improved in practical application. According to the different methods of information fusion, it can be subdivided into GIS-based AHP-type vulnerability index method, GIS-based ANN-type, GIS-based evidence-based and GIS-based weighted logical regression [79, 80]. This method has been widely promoted and applied in ore districts in north China and south China, playing an important role in solving the deep coal mining threatened by water disaster and in prolonging the service life of coal mine [81].

1.3 Main Content and Technical Approach

Based on the analysis of coal-forming age, water source and water filling method, according to the superposition relationship between the geological structure of the coal-bearing strata and the flow field characteristics of the regional karst water system, the geological structure patterns were divided into five types. The main factors influencing the floor water hazard were analyzed by employing both quantitative and qualitative analytic methods. The collection, quantification and dimensionless processing of these main control factors were systematically summarized, and the main control system of floor water inrush was established. An information fusion model of GIS-based JDL type water inrush evaluation model was built by applying information fusion theory and GIS technology. On this basis, a new and systematic method of water inrush evaluation from coal seam floor was established by using partition weighting theory, that is, the vulnerability index method based on partition weighting theory. Then this method was used to evaluate the risk of water inrush from coal seam floor of monoclinic sequence type #9 coal seam of Xiandewang Coal Mine and monoclinic reverse type #10 coal seam of Wangjialing coal mine. The main research contents are as follows (Fig. 1.1):

(1) Mine flooding hydro-geological conditions in north China Coalfield are analyzed. The sedimentary regularity of the main coal bearing formation-coal seam of Permo-Carboniferous is analyzed from the angle of coal-forming age, water source and flooding mode. The flooding characteristics of karst water in

Ordovician limestone, the most important water hazard threat in the north China coalfield, was studied and the water control role of regional structure, mining structure and the coal mine structure in north China coalfield was summed up.

(2) The geological structure relationships between Carboniferous Permian coal-bearing formation and underlying karst water system are studied. Based on superposition relationships between geologic structure of coal-bearing formations and flow field form of water filling source, the geological structure mode of karst water system and coal-bearing strata in north China major coal mines was divided into five different types: monoclinic sequence, monoclinic retrograde, parallel, synclinal basin, and fault block and others. The characteristics of various structural modes were studied respectively, and the general rule of floor water inrush from Ordovician limestone aquifer under the control of hydrogeological conditions was revealed, laying a foundation for establishing the hydrogeological conceptual model of the risk assessment of coal seam floor water inrush.

(3) Analysis of main control factors of water inrush from coal floor. Based on the analysis of the water diversion channel, water source and water filling strength of the coal seam floor, the paper analyzes the water inrush from the aquifer, the aquifer, the geological structure and the geological structure of the coal seam. Excavation engineering disturbance four aspects of the floor of the impact of water inrush analysis, and further study and summarizes the main control factor index collection, quantitative analysis and dimensionless processing methods. The main factor of water inrush from coal seam is established, which lays the foundation for multi—factor information fusion model.

(4) A new method for water inrush evaluation of coal seam floor. Firstly, the multi-source information fusion architecture and fusion algorithm of coal mine floor water inrush evaluation are studied by using information fusion technology. The data fusion model of JDL-based water inrush evaluation based on GIS is established. AHP and ANN are used to study the main control factors. The method of determining the weight. Based on the above two points of study, the author based on the method of vulnerability evaluation based on permanent rights. Based on the method of vulnerability evaluation based on permanent rights, the vulnerability index method based on partitioned variable weight is studied. This paper analyzes the method of constructing the zonal state vector of each main control factor, sums up the method of determining the variable range (threshold), transfer parameter and variable weight, and puts forward a new method to evaluate the water inrush risk of coal seam floor, Theoretical Vulnerability Index.

(5) Use the vulnerability index method based on the theory of partitioned variable weight to conduct vulnerability evaluation of water inrush in floor on monoclinic #9 coal seam and monoclinic inversed #10 coal seam of Xiandewang coal mine respectively. And the results of the evaluation are compared with the traditional water inrush coefficient method and the vulnerability index method based on the constant weight theory, respectively.

1.4 Innovation Points of the Research

(1) According to the superposition relationship between the geological structure of coal-bearing strata and the characteristics of regional karst groundwater flow field, five geological structure models in north China coalfields are proposed: monoclinic type, monoclinic inverse type, parallel strike type, syncline basin type, fault block and other types. Based on the study of floor water inrush from different structural modes, the general rule, under the control of water-filled hydrogeological conditions, of water bursting in Ordovician karst water of different structural modes is revealed.
(2) A method of collecting the main factors influencing the water inrush in coal seam floor and method of quantitatively analyzing these main factors are proposed. A quantitative index system of the water inrush evaluation of the coal floor is established, which lays the foundation for the water inrush evaluation.
(3) The information fusion model of multi–source main control factors influencing floor water inrush is put forward. Based on the theory of information fusion, GIS and modern mathematical methods, GIS-based JDL-type floor water bursting information fusion model is established.
(4) An improved method of evaluating water inrush risk of coal seam floor and steps how to implement this method are put forward. The method is the vulnerability index method based on partitioned variable weight theory. It breaks through the defects of the vulnerability index method based on the constant weight: the weight of the same main control factor in the evaluation process is a certain value and does not change with the change of the state value of its main control index. This improved method not only considers these main controlling factors and their corresponding weights, but also quantitatively determines the corresponding weights of the same factors in different state values, which solves one of the key technical problems in evaluation and prediction of water inrush vulnerability of coal seam.

1.5 Chapter Summary

(1) This chapter summarizes some research achievements of domestic and foreign scholars in the water inrush mechanism, water inrush prediction, and water hazard prevention and control technology.
(2) This paper focuses on: the coal-bearing formation of the Carboniferous Permian in north China coalfield and the geological structure mode of the underlying karst water; the new theory and method of evaluating the water hazard of the coal seam floor; taking the Xiandewang Coal Mine and Wangjialing Coal Mine as two examples to introduce and explain the improved evaluation method—vulnerability index method based on partitioned variable weight theory and its

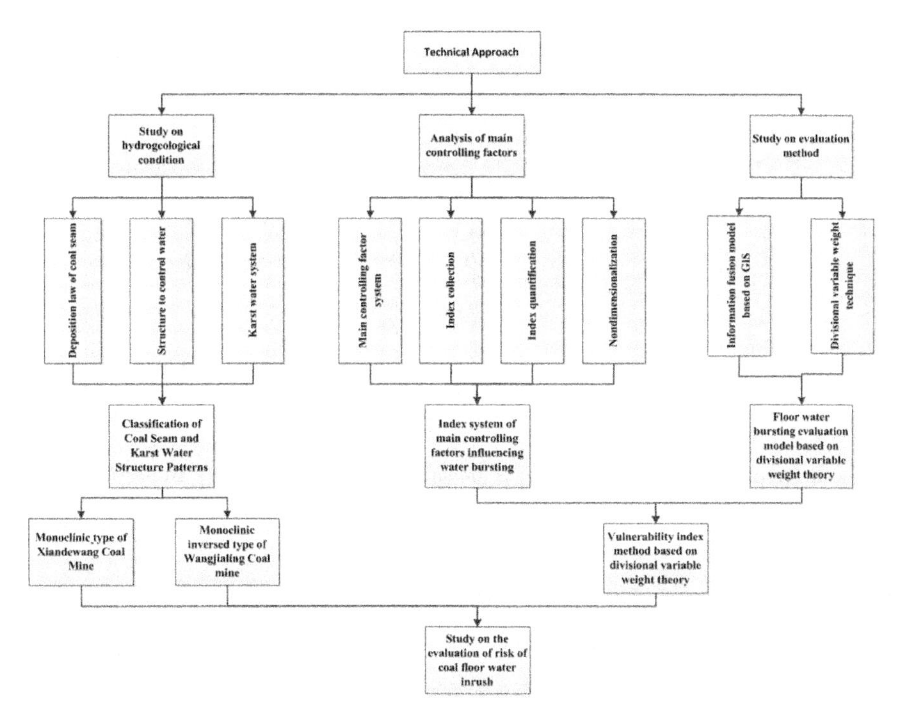

Fig. 1.1 Technical approach

application to the engineering application research of floor water inrush evaluation for ore districts with different structural model types.

References

1. Jiehua, Mao, and Xu Huilong. 1999. *Prediction and evaluation of coal resources in China*. Beijing: Science Publishing Company. (in Chinese).
2. Wu, Qiang, and Jin Yujie. 1995. *Mine water control decision system in North China Coal Mine*. Beijing: China Coal Industry Publishing House (in Chinese).
3. Gui, Herong, and Sun Benkui. 1999. Study meaning and the core content of the control theory about water inrush from bed bottom in deep mining. *Journal of Huainan Institute of Technology* 9(3): 211–220. (in Chinese).
4. Meng, Zhaoping, Gao Yanfa, and Lu Aihong. 2011. *Theory and method of water inrush risk assessment in coal mine*. Beijing: Science Publishing Company. (in Chinese).
5. Wang, Zuoyu, and Liu Hongquan. 1993. *Mining on confined water*. Beijing: China Coal Industry Publishing House 22–23,105–111. (in Chinese).
6. Organization and compilation of China Society of coal industry labor protection science and technology. 2007. *Mine water disaster prevention and control technology*. Beijing: China Coal Industry Publishing House. (in Chinese).
7. Xie, Heping. 2002. Present situation, basic scientific problems and prospect of resource exploitation under deep high stress. In *Frontiers and future of Science* (sixth episodes), 179–191,

ed. in chief of Xiangshan Science Conference, Beijing: China Environmental Science Press. (in Chinese).
8. Wu, Q., W. Zhou, J. Wang, and S. Xie. 2009. Prediction of groundwater inrush into coal mines from aquifers underlying the coal seams in China: Application of vulnerability index method to Zhangcun Coal Mine. *China Environmental Geology* 57 (5): 1187–1195.
9. Hu, Kuanrong, and Cao Yuqing. 1997. Basic theory research on water inrush and prevention principle of floor in mining face. *Journal Geol and Min Res North China* 12(3): 203–292. (in Chinese).
10. Fang, Peixian, Wei Zhongding, and Liao Zisheng. 2002. *Applied hydrogeology*. Beijing: Geological Publishing House. (in Chinese).
11. Wu, Qiang. 2002. *Mine flood control*. Xuzhou: China University of Mining and Technology press. (in Chinese).
12. Zhang, Jincai, and Liu Tianquan. 1991. *Simulated experimental study on stress and displacement of floor in coal mining face*. Beijing: Reference materials for coal scientific research. (in Chinese).
13. Qiang, Wu, and Wang Mingyu. 2006. Characterization of water bursting and discharge into underground mines with multi-layered groundwater flow systems in the North China Coal Basin. *Hydrogeology Journal* 4 (6): 882–893.
14. Wu, Q., Z. Zhang, and J. Ma. 2007. The new practical evaluation method of floor water inrush. I: The establishment of the main controlling index system. *Journal of China Coal Society* 32(1): 42–47.
15. Metallurgical Mine Design Institute. 1983. *The conditions of safe coal mining under water body and the development and practice of mine water prevention and control technology abroad*. (in Chinese).
16. Wang, Sujian. 2013. *Coal mine water disaster prevention and control technology research—Annual meeting of Shaanxi Coal Society*. Beijing: China Coal Industry Publishing House. (in Chinese).
17. Qiang, Wu, Li Wei, and Li Ruijun. 2008. Study on the assessment of mine environments. *Acta Geologica Sinica* 82(5): 1027–1034.
18. Wu, Qiang, and Xu Hua. 2004. On three-dimensional geological modeling and visualization. *Science in China* 47(8): 739–748.
19. Wu, Qiang, Xu Jianfang, Dong Donglin, et al. 2001. *Theory and method of geological disaster and water resources research based on GIS*, 129–130. Beijing: Geological Publishing House.
20. Wu, Qiang, and Liu Shouqiang. 2011. Discussion on prediction and evaluation of coal mine water disaster and emergency rescue plan. In *China Engineering Science and Technology Forum 118th field, 2011 International Coal Mine Gas Control and safety thesis collection*, ed. Xuzhou, 295–301, China, University of Mining and Technology press. (in Chinese).
21. Wei, Jiuchuan. 2000. *Study on fracture damage of floor rock and mechanism of floor water inrush*. Qingdao: Shandong University of Science and Technology. (in Chinese).
22. Slesarev, B. 1983. *The conditions of safe coal mining under water body and the development and practice of mine water control technology abroad*. China: Design Institute of metallurgical mines. (in Chinese).
23. Wolkersdorfer, C., and R. Bowell. 2004. Contemporary reviews of mine water studies in Europe. *Mine Water and the Environment* 23: 161.
24. Oda, M. 1986. An equivalent continuum model for coupled stress and fluid flow analysis in jointed rockmasses. *Water Resources Research* (13).
25. Sammarco, O. 1986. Spontaneous inrushes of water in underground mines. *International Journal of Mine Water* 5 (2): 29–42.
26. Bailey, W.R., J.J. Walsh, and T. Manzocchi. 2005. Fault populations, strain distribution and basement fault reactivation in the EastPennines Coalfield, UK. *Journal of Structural Geology* 27: 913–928.
27. Santos, C.F., and Z.T. Bieniawski. 1989. Floor design in underground coalmines. *Rock Mechanics and Rock Engineering* 22 (4): 249–271.

28. Sammarco, O. 1988. Inrush prevention in an underground mine. *International Journal of Mine Water* 7 (4): 43–52.
29. Crouch, S.L. 2005. Two-dimensional analysis of near-surface, single-seam extraction. *International Journal of Rock Mechanics and Mining Sciences and Geomechanics Abstracts* 10: 74–78.
30. Qiang, Wu, and Wanfang Zhou. 2008. Prediction of groundwater inrush into coal mines from aquifers underlying the coal seams in China: Vulnerability index method and its construction. *Environmental Geology* 56: 245–254.
31. Qiang, Wu, Wanfang Zhou, Guoying Pan, et al. 2009. Application of a discrete-continuum model to Karst Aquifers in North China. *Ground Water* 47 (3): 453–461.
32. Crouch, S.L. 2005. Two-dimensional analysis of near-surface, single-seam extraction. *International Journal of Rock Mechanics and Mining Sciences and Geomechanics Abstracts* (10): 74–78.
33. Dong, Q.H., R. Cai, and W.F. Yang. 2007. Simulation of water-resistance of a clay layer during mining: Analysis of a safe water head. *Journal of China University of Mining and Technology* 17(3): 345–348.
34. Kong, H.L., X.X. Miao, and L.Z Wang, et al. Analysis of the harmfulness of water-inrush from coal seam floor based on seepage instability theory. *Journal of China University of Mining and Technology* 17(4): 253–258.
35. Islam, M.R., and R. Shinjo. 2009. Mining-induced fault reactivation associated with the main conveyor belt roadway and safety of the Barapukuria Coal Mine in Bangladesh: Constraints from BEM simulations. *International Journal of Coal Geology* 79: 115–130.
36. Wu, Q., W. Zhou, G. Pan, and S. Ye. 2009. Application of a discretecontinuum model to karst aquifers in North China. *Ground Water* 47 (3): 453–461.
37. He, Zhihong. 2012. Application of groundwater inrush coefficient in safety mining evaluation. *Groundwater* 34(2): 35–36. (in Chinese).
38. Shi Longqing. 2012. Analysis of the origin of water inrush coefficient and its applicability. *Journal of Shandong University of Science and Technology* (NATURE SCIENCE) 31(6): 6–9. (in Chinese).
39. Qiang, Wu, Zhao Suqi, and Li Jingsheng. 2011. The preparation background and the main points of rule of mine prevention and cure water disaster. *Journal of China Coal Society* 36 (1): 70–74. (in Chinese).
40. Xu, Zhimin. 2010. *Mining-induced floor failure and the model, precursor and prevention of confined water inrush with high pressure in deep mining*. Xuzhou: China University of Mining and Technology. (in Chinese).
41. Wu, Q., and Y. Jin. 1995. *The decision system of water control in mines of north China type coal fields*, 45–51. Beijing: Coal Industry Public House of China.
42. Longqing, Shi. 2009. Review on mechanism of floor water inrush. *Journal of Shandong University of Science and Technology* 28 (3): 17–23. (in Chinese).
43. Jingming, Wang, Li Jingsheng, Gao Zhilian, et al. 1998. Coupling model of two phase flow in a fracture-rock matrix system and its stochastic feature analysis. *Journal of Coal Science and Engineering* 4 (1): 5–10.
44. Jingming, Wang. 1999. In-situ measurement and physical analogue on water inrush from coal floor induced by progressive intrusion of artesian water into protective aquiclude. *Chinese Journal of Geotechnical Engineering* 5 (21): 546–550. (in Chinese).
45. Zhang, Jincai. 1989. Theory and practice of prediction of water inrush from coal seam floor. *Coal Geology and Exploration* 4: 38–41. (in Chinese).
46. Zhang, Jincai, and Liu Tianquan. 1993. Analysis and research on mining factors of coal seam floor. *Journal of Coal Mining* 4:35–39. (in Chinese).
47. Zhang, Jincai, Xiao Kuiren. 1993. Study on mining failure characteristics of coal seam floor. *Coal Mining* (3): 44–49. (in Chinese).
48. Zhang, Jincai, Zhang Yuzhuo, and Liu Tianquan. 1997. *Seepage of rock mass and water inrush from coal seam floor*. Beijing: Geological Publishing House. (in Chinese).

49. Wang, Zuoyu, and Liu Hongquan. 1993. *Water mining on confined water*, 130. Beijing: China Coal Industry Publishing House. (in Chinese).
50. Wang, Zuoyu, and Liu Hongquan. 1989. Study on water inrush mechanism of coal seam floor. *Coal Geology and Exploration* (1): 11–13. (in Chinese).
51. Institute of geology, Chinese Academy of Sciences. 1992. *Study on water inrush mechanism of karst water in coal mine of China*. Beijing: Science Publishing Company. (in Chinese).
52. Wang, Zuoyu, Liu Hongquan, and Wang Peiyi, et al. *Journal of China Coal Society* 19(1): 40–48. (in Chinese).
53. Zhang, Jincai. 1998. *Study on failure and seepage characteristics of mining rock mass*. China: General Research Institute of coal science. (in Chinese).
54. Qian, Minggao, Shi Pingwu, and Xu Jialin, et al. 2010. *Mining pressure and strata control*. Xuzhou: China University of Mining and Technology press. (in Chinese).
55. Wu, Q., H. Xu, and W. Pang. 2008. GIS and ANN coupling model: An innovative approach to evaluate vulnerability of karst water inrush in coalmines of north China. *Environmental Geology* 54 (5): 937–943.
56. Hongge, Ni, and Luo Guoyu. 2000. Analysis of water control and stability control mechanism of dominant face in underground mining. *Journal of Engineering Geology* 08 (03): 316–319. (in Chinese).
57. Minggao, Qian, Liao Xiexing, and Li Liangjie. 1995. Mechanism of the fracture behaviours of mainfloor in longwall mining. *Chinese Journal of Geoteehnieal Engineering* 17 (6): 56–61. (in Chinese).
58. Li, Liangjie, Qian Minggao, and Wen Quan, et al. 1995. Relationship between the stability of floor structure and water-inrush from floor. *Journal of China University of Mining and Teehnology* 24(4): 18–23. (in Chinese).
59. Li, Liangjie. 1995. *Study on mechanism of water inrush from mining floor*. Xuzhou: China University of Mining and Technology press. (in Chinese).
60. Qian, Minggao, Liao Xiexing, and Xu Jialin, et al. 2003. *Key stratum theory of strata control*. Xuzhou: China University of Mining and Technology press. (in Chinese).
61. Xu, Jialin, and Qian Minggao. 2004. Study and application of mining-induced fracture distribution in green mining. *Journal of China University of Mining and Technology* 33(2): 141–144.
62. Pu, Hai. 2007. *Water retaining key stratum model and mechanical analysis and application of water preserved coal mining*. Xuzhou: China University of Mining and Technology press. (in Chinese).
63. Wu, Q., and W. Zhou. 2008. Prediction of groundwater inrush into coal mines from aquifers underlying the coal seams in China: Vulnerability index method and its construction. *Environmental Geology* 55 (4): 245–254.
64. Hongge, Ni, and Luo Guoyu. 2000. Study on the mechanism for water transportation and rock-mass stability controlled by pereferred plane in underground mining. *Journal of Engineering Geology* 08 (03): 316–319. (in Chinese).
65. Li, Dinglong, Wang Moling, and Zhou Zhian. 1997. Multi-resources information method and its application to prediction of water-burstinginmine. *Journal of Catastrophology* 12(3): 316–319. (in Chinese).
66. Wang, Shuyuan. 1989. Fuzzy mathematics prediction method of mine water inrush events. *Journal of Shandong Mining Institute* 8(3): 48–51. (in Chinese).
67. Xu, Yanchun, and Geng Deyong. 1992. Fuzzy cluster analysis and prediction of shaft lining failure. *Journal of Coal Science and Technology*, 1992 (7): 16–19. (in Chinese).
68. Li, Fuping. 1997. Discussion On Prediction Method Of Water Inrush In Coal Mining Face. *Hebei Coal Journal* (2): 8–10. (in Chinese).
69. Li, Dinglong. 1998. Application of neural network to prediction and prediction of coal mine water inrush. *Coal Journal* 7(6):1 8–19. (in Chinese).
70. Wu, Q., Y. Liu, and L. Yang. 2011. Using the vulnerable index method to assess the likelihood of a water inrush through the floor of a multi-seam coal mine in China. *Mine Water and the Environment* 30 (1): 54–61.

71. Yanchun, Xu. 1996. Macroscopic prediction of mine water disaster with grey theory. *Journal of Coal Mining* 1: 46–48. (in Chinese).
72. Sun, Yajun. 1989. *Application of information fitting method to prediction of floor water bursting in eastern Jiaozuo mining area*. Xuzhou: China University of Mining and Technology. (in Chinese).
73. Zheng, Shishu, and Sun Yajun, et al. 1994. Application of GIS to prediction of face water inrush under the Weishanhu lake in Yinzhuang Coal Mine. *Journal of China University of Mining and Technology* 23(2): 48–56. (in Chinese).
74. Zhang, Dashun, Zheng Shishu, and Sun Yajun, et al. 1994. *Geographic information system and its application in coal mine water disaster prediction*, 108–169. Xuzhou: China University of Mining and Technology press. (in Chinese).
75. Zhong, Guosheng, Jinag Wenwu, and Xu Guoyuan. 2007. Prediction of water inrush from floor based on catastrophe theory. *Journal of Liaoning Technical University* 26(2): 216–218. (in Chinese).
76. Bai, Chenguang, Li Liangjie, and Yu Xuefu. 1997. Cusp catastrophe model for instability of key stratum of confined water slab. *Journal of China Coal Society* 22(2): 149–154. (in Chinese).
77. Wu, Qiang, Xu Jianfang, Dong Donglin, et al. 2001. *Theory and method of geological disaster and water resources research based on GIS*, 129–130. Beijing: Geological Publishing House. (in Chinese).
78. Wu, Qiang, Zhang Zhilong, and Zhang Shengyua. 2007. A new practical methodology of the coal floor water bursting evaluating II—The vulnerable index method. *Journal of China Coal Society* 32(11): 1121–1126. (in Chinese).
79. Qiang, Wu, Chen Peipei, and Dong Donglin. 2002. Hazard assessment system for ground fissure based on coupling of ANN and GIS—A case study of ground fissures in Yuci City, Shanxi Province. *Seismology and Geology* 6: 249–257. (in Chinese).
80. Wu, Qiang, Xie Shuhan, and Peizhenjiang, et al. 2007. A new practical methodology of the coal floor water bursting evaluating III—The application of ANN vulnerable index method based on GIS. *Journal of China Coal Society* 32(12): 1301–1306. (in Chinese).
81. Wu, Qiang, Zhang Zhilong, Ma Jifu. 2007. A new practical methodology of the coal floor water bursting evaluating I—The master controlling index system construction. *Journal of China Coal Society* 32(1): 42–47. (in Chinese).

Chapter 2
Structural Pattern and Characteristics of Floor Rock's Water-Soluble and Flushing Water System

2.1 Sedimentation Characteristic in Coalfields of North China

The north China-type coalfield refers to a set of Permo-Carboniferous sea-land interacted coal-bearing rock system which is not integrated in the Ordovician (or Cambrian) thick layer of carbonate rocks after the transformations of subsequent structure. Permo-Carboniferous coal-bearing strata in northern China is very broad distributed, the northern boundary is the Yin mountains, Yan mountains archicontinent and Shenyang and the uplifted eastern section, the southern boundary is the Qinling Mountains, Funiu Mountain and the Dabie mountains archicontinent, east to the Jiaoliao archicontinent and is adjacent to the Yellow Sea, the Bohai Sea, west to Helan Mountain- Liupanshan structural belt, which forms a huge north China coal gathering area crossed 15 provinces, cities and districts [1]. Including the whole area of 4 provinces and cities of Hebei, Shanxi, Henan, Shandong, Beijing, Tianjin, southern part pf Jilin, Liaoning, Inner Mongolia provinces, northwestern part of Jiangsu Province, northern part Anhui and Shaanxi provinces, and eastern part of Gansu and Ningxia provinces (Fig. 2.1).

The formation of the north China coalfields is determined by its special crustal tectonic activity and seawater advance and retreat. As early as the Proterozoic, before the formation of north China coalfields, the North China area as an uplift area consists of a set of metamorphic rocks with strong stiffness and integrity. From the Cambrian to the end of the Permian, the crustal movement in north China is mainly caused by the fault block wane's lifting activities, and the magmatic activities are also not obvious. From the Cambrian to the Ordovician, the north China platform slowly sank. The seawater invaded from south to north, from east to west and formed a Cambrian-Ordovician shallow marine carbonate formations. From the Silurian to the Lower Carboniferous, under the influence of the Caledonian movement, the whole north China platform slowly rose to land. The long time weathered erosion, almost caused the complete missing of upper Ordovician to lower Carboniferous strata, and even part areas' the middle and lower Ordovician were also missed, the Middle Carboniferous was not directly integrated on the Cambrian system [2, 3].

Y. Zeng, *Research on Risk Evaluation Methods of Groundwater Bursting from Aquifers Underlying Coal Seams and Applications to Coalfields of North China*, Springer Theses, https://doi.org/10.1007/978-3-319-79029-9_2

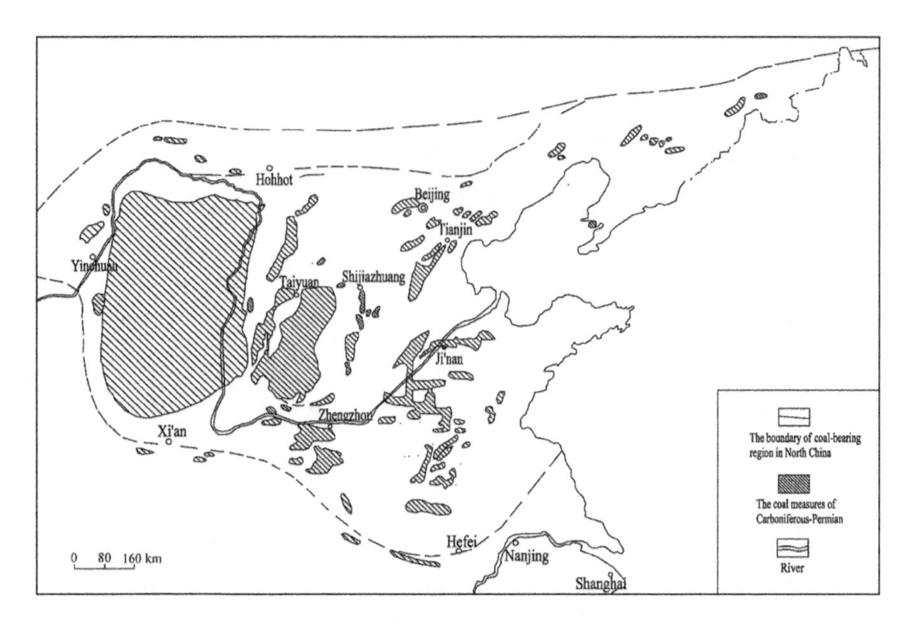

Fig. 2.1 Distribution of coalfields in north China

From the beginning of the Middle Carboniferous, the north China platform began to sink slowly again, and the seawater invaded again from south to north, from east to west, thus deposited a set of land-sea alternated Benxi formation and Taixi formation coal-bearing stratum in the vast Ordovician limestone weathering surface. From the Late Carboniferous to the Permian, the north China platform uplifted again and began to gradually regress from north to south and from west to east. In this slow process of regression, the Shanxi formation, upper series and the lower series were extensively and continuously deposited. During the seawater's "invade, retreat, and invade again, and back again" period, it formed the famous Carboniferous Permian land-sea alternated phase of coal construction, which is, north China coalfields. The total thickness of the Carboniferous Permian coal-bearing strata in north China is generally between 700 and 1000 m.

2.1.1 Main Deposition Law of the Carboniferous Coal-Bearing Rock Series

The Benxi Formation is not integrated in the ancient weathering crust of Ordovician limestone, and is mainly composed of the sediments of coastal, shallow sea sandstone, siltstone and mudstone, and several limestone layers. Due to in the early stage of deposition the paleogeography is undulated, and the sedimentary thickness changed greatly. It is the thickest in the eastern part of Liaoning which can reach to 150 m,

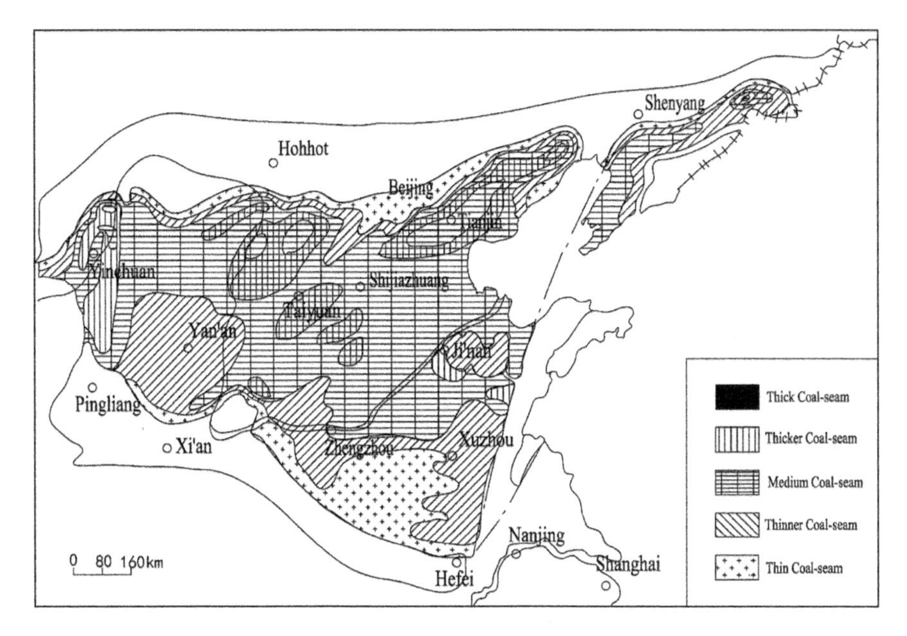

Fig. 2.2 Total thickness of coal seams in the late Carboniferous Taiyuan group in north China

and the thinnest part of the deposition is only tens of meters. Its deposit thickness has the changing trend of north thick south thin, east thick west thin (Fig. 2.2). The coal seams contained in the Benxi Formation are generally located in the middle and upper part of the Benxi Formation, most are thin seams and seam lines, and some areas also sandwich seam lines in the lower part. Except the northern part of northeast China and Tangshan having abundant coal seams and depositing with the minable seam, the other areas only have 1–2 layers of coal line, and their minable coal is very little.

The Taiyuan Formation is mainly composed of fine-grained sandstone, siltstone, mudstone, limestone and coal seam [4]. The coal facies are mainly sea-land transitional facies, shallow sea facies and swamp facies. The number of layer and thickness of thin limestone are decreasing and attenuating from south to north, and the proportion of clastic rocks is gradually increasing from south to north. Among them, Kaiping, Xinglong coal fields which are in north of Shijiazhuang and coal field in northeast are mainly composed of clastic sandstone.

The Taiyuan Formation is land—sea alternated coal—bearing rock series and is one of the main coal-bearing strata in the whole region. The number and thickness of the coal seam are very different in the south and north, and the number and thickness of the coal seam in the north are more and thicker than that in the south. It always forms thick and super thick coal seams whose total thickness always more than 10 m. The thickness of the coal seam in Shanxi Pingshuo ore district is approximately 30 m, and it reduces gradually as it is to south. In the Xinmi coal field, Huainan coal field and Huaibei coal field in southern part, the total thickness of coal seam is generally less than 5 m, mostly are thin coal seam or very thin coal seam [5].

2.1.2 Main Sedimentary Regularity of Permian Coal-Bearing Rock Series

North China-type coalfield from the later period of Late Carboniferous, the sea water exited to the southeast, and the coastline also gradually migrated to the southeast. Therefore, the Shanxi Formation of the Permian and the upper and lower series are mainly composed of continental sediments and lack of marine limestone deposits.

The Shanxi Formation is also one of the main coal-bearing strata in the north China Coalfield. Its lithology is based on sandstone, siltstone and mudstone, the interbedded coal seam exclude limestone. The main lithofacies are land facies and the transition facies, the total sedimentary thickness ranges from 40 to 140 m, and the total trend of sedimentary rock thickness is also the same as that of Taiyuan Formation (Fig. 2.3). In the period of early Permian coal-accumulating is still at its peak, all the north China have coal seams occurrence [6]. Along south of the Yanshan archicontinent the Helan Mountain, Dongsheng, Datong, west of Beijing, Nanpiao and other coalfields have thick coal seam to form, the thickness of southward coal seam gradually get thinned, the total thickness of the northern coal seam can be 3–30 m, the central is 2–10 m, south is 0–7 m. overall, Shanxi coal seam distributed in a wide range in the whole of north China coal field, and relatively more uniform. But its coal bearing property still shows the law of descending from north to south. The coal strata contained in the upper and lower series are almost free from deposition in the north of Xi'an and north of Zhengzhou, and are generally not treated as coal-bearing rock systems. To the late period of late Permian, the climate has been converted to dry, and the coal-accumulation process in late Paleozoic also ended.

2.2 Hydrogeological Features and Characteristics of Water-Filled Coalfields in North China

The general trend of the Late Paleozoic seawater's advance and retreat in the north China platform is that the seawater invaded from the east to the west in the Middle Carboniferous, and in the late Carboniferous it gradually retreated to the southeast again. In the early Permian, the seawater has completely withdrawn from the area, and the north China platform rose to land completely. The number and thickness of limestone in coal-bearing strata are also gradually reduced from bottom to top, from southeast to northwest. The changing characteristics of this ancient geography determines that there are obvious regulations of plane divisional and profile zone in hydrogeological conditions of the coalfield in this area [6, 7].

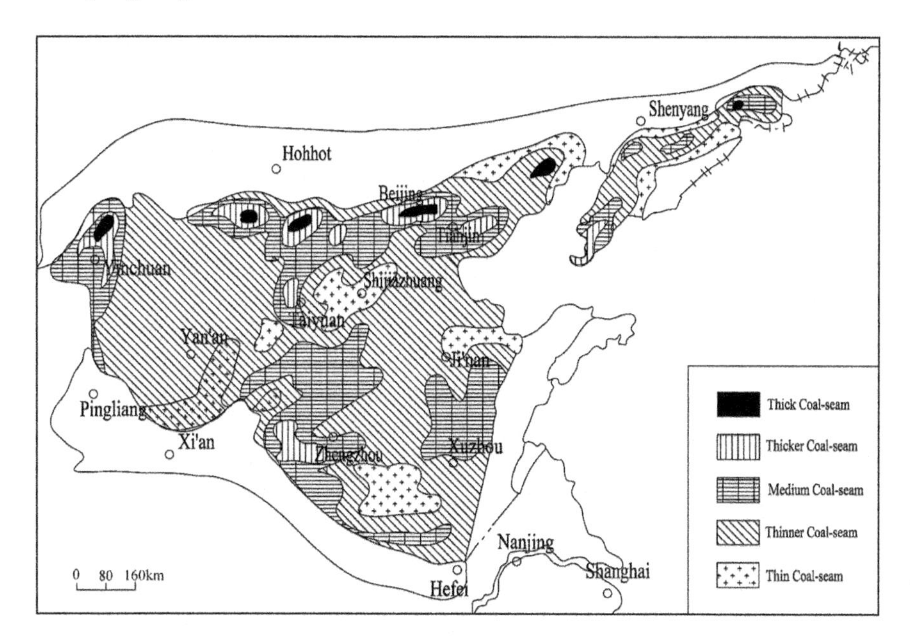

Fig. 2.3 Total coal seam thickness early Permian Shanxi group in north China

2.2.1 Hydrogeological Features

The aquifer group in the north China coalfields has the characteristics of heterogeneity, anisotropy, irregular spatial distribution and complex boundary conditions. In the plane, it can be divided into the southern, central and northern three areas:

[1] The Huainan, Huaibei and Qishan areas, in the late Carboniferous, are mainly the coastal—shadow sea environment. The thickness of the sedimentary cycle is small but the number of it is high. The limestone has high proportion, accounting for 60% of the total thickness of the Taiyuan Formation. The coal seams have a large quantity but are also flimsy, and most of them are unacceptable. The main coal-bearing strata is Shanxi formation and Shihezi series, mining seam is very far from the Austrian gray aquifer. Austrian gray water generally cannot directly threaten coal mining. Its main hydrogeological problems, the first one is the thick Cenozoic aquifer directly cover on the coal outcrop, the second is the Taiyuan Formation's aquifer of thin layer of limestone and its connection with the Austrian gray water level. Hydrogeological conditions are generally not very complicated.

[2] The north of above areas to the southern area of Taiyuan, Shijiazhuang line, in the late Carboniferous, are the coastal plains coal-gathering environment, Taiyuan Formation contains the mainly available coal seam. There is just a Benxi Formation of 20–60 m between the lower coal and lower Ordovician, and Taiyuan Formation and Benxi Formation all contain multi-layer limestone,

and have close contacts with the lower Ordovician aquifer. In the coal mining, Ordovician limestone water and limestone water of Taiyuan Formation and Benxi Formation are very easy to inrush into the mine, which often cause flooding accidents and make it difficult to develop a large amount of coal resources in the Taiyuan Formation. This area's coalfield hydrogeological conditions are the most complex one in north China coalfields.

[3] The area of Taiyuan, Shijiazhuang and north of this region has thicker Benxi formation, and less limestone. The limestone in Taiyuan Formation is also less and thin, or even has no limestone. Taiyuan Formation and Benxi Formation themselves do not have the problem of karst water. The underlying Ordovician is generally not easy to directly threaten the mine, so coal mine floor water inrush and the times of mine inundation in the region are less than the central region of north China, which is south of this region. In general, the coalfield hydrogeological conditions of this area are relatively simple compared to the central part of north China.

2.2.2 *Hydrogeological Features in Profiles*

2.2.2.1 Middle Ordovician Karst Aquifer Group

The middle Ordovician soluble rock which is based on the limestone has wide distribution area and with large thicknesses, and has gypsum horizon, its karst development is strong and is the most important karst aquifers in northern China. The middle Ordovician karst aquifer group is the basement of coal-bearing stratum with strong water abundance and is also the main aquifer in northern coalbed and the main threat of the developing of this area's coal field.

The Carboniferous Permian coal-bearing strata in this area in generally not integrated with the middle Ordovician limestone. The Ordovician and Cambrian limestone have experienced long-term sedimentation from the Upper Ordovician to Lower Carboniferous before the coal deposition. And the karst develops generally among which the development of middle Ordovician limestone karst is more intense [8]. In the later movements of Indosinian, Yanshan and Himalayan, the karst development in the uplift and tectonic damage areas is further exacerbated. In the uplift area, the limestone is exposed, and the recharge conditions of karst water are further enhanced. Therefore, the basement of north China carboniferous Permian coalfield has the karst aquifer with abundant water content and huge water structure. In exploiting coal seam, the water in the lower high-pressure karst aquifers often break through the floor of ore district or roadway, or by means of fault, karst collapse column and other water conduits to flood into mine and cause waterlogging disaster.

Due to the different thickness of the original deposition and the late weathering, the thickness of the middle Ordovician in the overall shows the characteristic of thick in middle, and thin in the north and south side. In the middle part of the line of Shijiazhuang, Yangquan, the Ordovician thickness could up to 640–784 m and gradually become thin. North to the Junggar coal field in Yinshan archicontinent, south to Luoyang, Yuxian in Dabie Mountains archicontinent, it is all missing. The coal base is Ordovician dolomitic limestone or Cambrian carbonate rocks [9].

In according to the sedimentary circle, middle Ordovician can be divided into Fengfeng formation, upper and lower Majiagou formation. According to the regional lithology, lithofacies and fossil combination, each formation is divided into 2–3 segments. Each formation's upper part is the thick limestone with great permeability, the bottom part is the less permeable brecciated limestone.

The base of the coal measures is different from that of the Middle Ordovician. There are also significant differences in the hydrogeological conditions between them. In the Fengfeng, Jincheng and other ore districts, coal measure basement directly contact with Fengfeng formation. When exploit Taiyuan Formation coal seams, Fengfeng formation aquifer can be directly contact with Taiyuan coal seam floor and is the direct water—filled aquifer of coal seam floor water-inrush; but the water-bearing formation of upper and lower Majiagou formation can only by the aid of inner boundary to indirectly fill water to coal mine and to be the indirect water filling aquifer of coal floor water inrush. In Jiaozuo coal ore district of Henan province, the Fengfeng formation is in direct contact with Majiagou formation, so the upper Majiagou formation can be the direct water filling aquifer and the lower Majiagou formation is still the indirect water filling aquifer. The ore districts like Yuxian, Dengfeng and Pingdingshan are lack Ordovician, and their Permo-carboniferous system directly undulate on the Cambrian, who has weaker karst degree and worse water abundance. Therefore, the coal field in Yuhuaiaoxian has simpler hydrogeological conditions.

The coal-bearing strata limestone aquifer group is the second largest threat of water damage in this area. The Carboniferous sea-land interaction coal-bearing strata, contains 3–11 layers of thin layer of limestone with the total thickness of generally 10–40 m, and it is the pressure karst fissure aquifer. However, due to thin layer of limestone aquifer has the limitation, and often be cut into a variety of fault blocks with limited scale by fault, so its impact of the seriousness on the deposit water filling depends on whether get the supply of the lower Ordovician limestone water or shallow subtle outcrops Quaternary strong aquifer [10]. When they have no hydraulic connection, the source of the deposit water filling is mainly the consumption of the thin layer of limestone static reserves, and they do not have much threat on safety of mine.

Taiyuan Formation is the most complex coal-bearing strata in this area. It is mainly composed of confined karst fissure aquifers and the Middle Ordovician karst aquifers is the nearest one beside it, which is only 20–60 m of Benxi Formation between them. In Luzhong and Jizhong area, Benxi Formation also contains Xujiazhuang limestone with high water abundance and Caopugou limestone, which increased the threat in mining coal seams roadways in Taiyuan Formation. Taiyuan Formation is sea-land alternated deposition, its coal measures contain multi-layer limestone and has the characteristic of coal rock interbedding. These limestone layers have karstification and water abundance, and often have close hydraulic connection with the underlying

Austrian gray aquifer. When exploit Taiyuan coal seam, the limestone will become an aquifer which directly water fill the mine roadway [11].

Shanxi Formation is basically the continental deposition, has no limestone and small water volume, and is mainly composed of sandstone fissure aquifer. However, due to the thin thickness of the aquiclude between the coal seam and the limestone aquifer, when the water pressure in the lower limestone aquifer is high, or the existence of fracture, karst collapse column and other water conducting inner boundary, it results in the underlying karst water swarming into the mine in a large quantity. But in general, its hydrogeological conditions are relatively simpler than that of the Taiyuan Formation.

The upper and lower series are all the continental coal bearing strata and have long distance between the underlying limestone. It does not exit the problem of water-inrush of karst [12]. The main water filling source is sandstone fissure water in small water volume. The hydrogeologic conditions of coal field are relatively simple.

2.2.2.2 Unconsolidated Porous Aquifer Formation

It is mainly in the Quaternary alluvial-pluvial sand layer, gravel layer, sand cobble layer other sediments. It is also distributed in the alluvial-pluvial plains in the southeastern foot of Taihang mountains, southern foot of Yan mountains, middle or southern part of Shandong, northern part of Anhui and other areas; Liaohe Plain, Haihe Plain, the Yellow River Plain and the Huaihe Plain; Taitong basin on the west side of Taihang mountains, Xinzhou Basin, Taiyuan Basin and other rift basins. It has different water storage capacity in different areas, the buried depth of aquifer in the alluvial plain is approximately 50 m, the thickness of it is 5–25 m, and the water burst of one unit is generally 192–480 m^3/d. It buried shallowly in Piedmont plain for approximately 5–10 m, thickness of it is 30–50 m, and the water burst of one unit is 8.3–16 L/s/m. The thickness of the aquifer in the rift basin is 10–50 m, and the buried depth in various basins is different. And, the water burst of one unit is 0.5–16.7 L/s/m. In the Kailuan, Huainan, Huaibei, Yanzhou and other ore districts, there are Quaternary cover layers with the thickness of more hundreds of meters, and these thick loose porous aquifer formations are not integrated with the coal-bearing strata and Ordovician limestone [13]. The Quaternary bottom clay aquifer's depositional thickness of these above areas is an important factor to determine the complexity of the hydrogeological conditions of the deposit. Once the clay aquiclude is thinner or partially absent, the loose pore aquifer formation will become one of the main water sources of mine's water-filling.

2.2.2.3 Aquiclude Formation

Controlled by the paleogeographic conditions of the Carboniferous, the general trend of Benxi Formation's deposit thickness is east thick and west thin. Benxi Formation from bottom to top can be divided into three parts: the lower part is the purple mudstone with a thickness of 15 m; the middle part is yellow sandstone, sandy limestone clip lens limestone with 4–10 layers' coal line and the thickness of approximately 75 m; the upper part is the yellow mudstone, fine sandstone limestone clip and is approximately 55 m.

The hydrodynamic lithology of Benxi Formation plays an important role in the hydrogeological conditions of the north China coalfield. At the same time, the middle part of the Benxi Formation also contains a multi-layer limestone, and the karst degree and water abundance but also play a direct role in the mine water-filling. The number and thickness of limestone in the middle of the Benxi Formation are gradually increasing from west to east.

Overall, the Benxi Formation is a relatively aquiclude formation, which is conducive to the exploitation of the Carboniferous Permian coal seam. When the thickness and intensity of lithology of the Benxi Formation is sufficient to resist the water pressure of the Ordovician limestone water, it is difficult for Ordovician limestone to directly swarm into the mine. It can only by means of fault, the collapse of the column and other inner boundary to fill water to the mine, or through the fault to supply the Carboniferous limestone aquifer, and thus become an indirect water-filled aquifer. Therefore, many coal mines in the east and north of the north China coalfield can safely exploit the coal seams of Taiyuan Formation. But when the mining depth is deeper, the Ordovician limestone's water pressure is higher, and the thickness of the Benxi Formation is thin, the Ordovician limestone water will break through the overlying aquiclude and enter the coal seam of Taiyuan Formation and even into the mine of exploiting Shanxi Formation coal seam. Xingtai, Fengfeng, Hebi, Jiaozuo that in the eastern and southern foot of Taihang mountains and Xinggong in Henan, Huoxian in Shanxi and other coal fields are the same. In many mines or sections of these ore districts, measures such as drainage and depressurization must be taken so that it does not exceed the compressive capacity of the Benxi Formation or take other effective control measures to safely mine the Taiyuan coal seam.

2.2.3 *Tectonic Reconstruction on Hydrogeological Conditions of Coal Fields*

In the tectonic rising area, it makes Ordovician limestone and the Carboniferous limestone exposed to the surface, and accept the supply of rainfall and the surface water. The groundwater circulates intensively and the degree of karstification is further exacerbated, which is unfavorable to the exploitation of shallow and middle deep coal seams. In the tectonic descent area, due to poor supply and discharge conditions, long-term stagnation of groundwater, the karstification is difficult to develop. In the long geological age, the existing karst pores, cracks were filled by high salinity sediment in the groundwater and roof rock slabs, so that the degree of karst greatly reduced or even "cured." These coal floors have high water pressure, but the amount of water inflow is smaller [14].

The ancient karst system, which is relatively unified from west to east, is divided and blocked by tectonic action. The ancient karst system is transformed into many new karst systems and subsystems which are isolated or semi-isolated. Each karst system has its own supply and discharge areas, respective supply and discharge methods. There are scattered groundwater runoff and concentrated strong runoff. The karst has both a wide distributed karst pores, gaps and karst cave and underground river. In short, the post-tectonic action makes the karst degree in area more intense, and the coalfield hydrogeological conditions more complicated.

2.3 Main Features of Floor Rock's Water-Soluble and Flushing Water System in North China Coalfields

The karst water system occupies a very important position in the floor water filling and water supply of ore district in the north China coalfields. The karst water-filled deposits are widely distributed, and the exploitation of coal in the middle and deep part of the north China coalfields is universally faced with the problem of karst water filling. The general hydrogeological conditions of karst water-filled deposits are more complicated. Water-yield and water-inrush happened frequently. And, water flow is of a large amount. Flood accident happened frequently in this system.

2.3.1 Coexistence of Coal and Water

In early Paleozoic, the north China platform widely deposited the most important karst water-filled aquifer: Cambrian Ordovician carbonate water-filled aquifer formation. The Caledonian movement raised the whole area to a land and made this area's widely distributed Cambrian and Ordovician suffer from weathering [15]. As to the Carboniferous, it has entered the coal-forming phase of the sea-land interaction, and deposited the Carboniferous Permian coal-bearing strata, makes the coal-bearing rock series directly contact with the Ordovician carbonate rocks. The vertical distance between the lower coal group and the underlying Ordovician karst aquifer is 20–60 m (Fig. 2.4). The water conduits like fissures formed in the process of linear structure fissures, point-like collapse column and coal mining generate direct and indirect hydraulic connection between karst water and groundwater, and constituted the system of "coexistence of water and coal".

According to statistics from Liang Yongping et al., karst water in north China can be divided into 119 karst water systems, most of which contain coal-bearing strata, and 52 karst water systems are distributed with coal fields, a total of 47 ore districts. According to the distribution law of karst water in north China and the conditions of fill, diameter and row, and the distribution of the main ore districts in north China, we can roughly determine the location of ore district in the karst water system, and

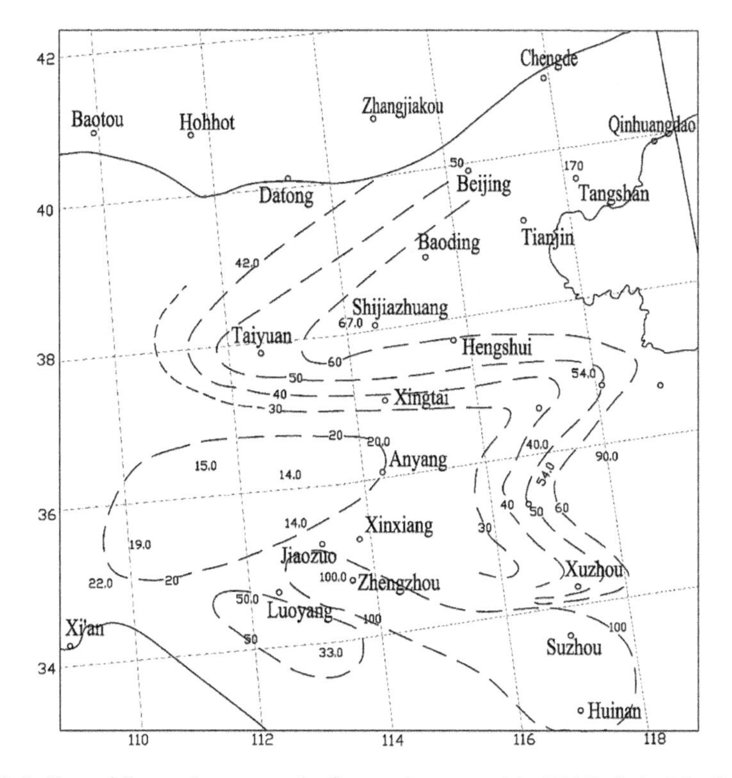

Fig. 2.4 Isolines of distance between major floor coal seams and the Middle Ordovician limestone

according to the flow field of karst water system, roughly analyze the features of karst water—filled aquifer in ore district (Table 2.1).

2.3.2 Three-Dimensional Recharge Characteristics

Due to the large scale and high openness of the karst water system in the north, there is a unified water cycle in the system formed by atmospheric precipitation, surface water, old air-water, loose layer pore water, clastic fissure water, karst fissure water and karst water. As a basement of coal-bearing rock, The Middle Ordovician limestone aquifer group has a thickness of approximately 200–800 m, abundant water supply and strong water enrichment. The Carboniferous sea-land interacted coal-bearing strata that integrated with Middle Ordovician or Cambrian giant layer of carbonate rocks, containing 3–11 layers of thin layered limestone; the total thickness is generally 10–40 m, is pressure karst fissure aquifer. Part of the north China coalfield, Quaternary aquifer group is not integrated over the coal and the Austrian gray above, such as the Kailuan Donghuantuo Mine. Quaternary bottom clay aquiclude's

Table 2.1 Characteristics of karst water systems in major mine areas in north China inundated with karst water

Ore district	System name of karst water	Location of the mine in the system		
		Upstream of runoff area and retention area	Confined area in the midstream of runoff area	Runoff discharge confined area
Kailuan	Kaiping syncline	Zhaoge village, Tangjia village, Lvjiatuo	Tangshan, Jingezhuang, Majiagou	Fangezhuang
	Chezhoushansyncline		Donghuantuo	
	Xiacang—Linnancang		Linnancang	
Xishan	Lancun-Jinsi spring	Malan, Duerping	Ximing, Xiyu, Dongshan	
Huoxian	Guozhuang spring	Gaoyang, Shuiyu	Zhangjiazhuang, Fujiatan	Bailong, Tuanbo
Yangquan	Niangziguan spring	First, Second, Third coal	Fourth coal, Yinying	
Jingjin	Weizhou spring	First, third coal		
Luan	Xian spring	Changcun	Wuyang, Wangzhuang, Shigejie	
Jincheng	Sangu spring	Yingshan, Niushan	Fenghuangshan, Gushuyan	
Xingtai	Baiquan	Mowo, Dongpang	Zhangcun, Xiandewang, Guoerzhuang	Gequan, Xingtai
Fengfeng, Handan	Heilong dongquan	Yangquhe, Jiulongkou	Second, Second, Fourth, Fifth coal Wangfeng, Tonger, Xuecun	Third coal, Sunzhuang, Huangsha
Anyang	Pearl spring		Tongzhi	Zizhen
Hebi	Xujiagou spring	First, Second, Forth coal	Third, Fifth, Sixth coal	Eighth coal
Jiaozuo	Jiulishan spring		Wangfeng, Zhucun	Xiaoma, Zhongma, Jiulishan
Yanlong	Fault block in Yanlong	Shangzhuang	Jiaocun, Liuzhuang	Zhuge
Ronggong	Fault block in Ronggong	Dagouyu, Xinzhong	Wanghe, Xuzhuang	
Xinmi	Fault block in Xinmi		Zhucun, Lugou	Dongfeng
Zibo	Zibo syncline	Xihe, Nanding, Shigu	Hongshan	XIazhuang, Longquan
XInwen	Xinwen	Quangou	Liangzhuang, Sun village, Zhangzhuang	Ezhuang

(continued)

Table 2.1 (continued)

Ore district	System name of karst water	Location of the mine in the system		
		Upstream of runoff area and retention area	Confined area in the midstream of runoff area	Runoff discharge confined area
Feicheng	Feicheng	Dafeng, Caozhuang, Yangzhuang	Southern Gaoyu, Taoyang	
Yanzhou	Yanzhou	Xinglongzhuang, Nantun, Beisu		
Zaozhuang	Taozao	Taozhuang, Tiantun	Shanjialin, Zhuzigao	
Hancheng	Hancheng	Magouqu, Xiangshan	Sangshuping	
Pubai	Tongpu	Xiping	Baishui, Macun, Quanjiahe	Wangcun, Heyang
Xuzhou	Xuzhou composite fold	Zhangji, Zhangxiaolou, QIshan	Xinhe, Hanqiao, Quantai	
Huaibei	Huaibei composite fold	Shenzhuang, Yuanzhuang	Shuoli, Daihe, Zhuxianzhuang	Xiangcheng, Yangzhuang
Huainan	Huainan	The first mine on Panji	Kongji, Lizuizi, Xinzhuangzi	Liyi coal, Lier coal

deposition is very thin, part area even completely missed deposition. Loose pores aquifer directly covers on the concealed outcrops of aquifer and the middle Ordovician limestone aquifer, and form another major source of water in the ore district. And there is a direct or indirect transformation relationship between the various types of groundwater, the hydraulic connection is close, and formed a special deposit water hydrogeological conditions.

The main aquifer group in the north China coalfields are the middle Ordovician karst aquifer group, the Taiyuan Formation under pressure karst fissure aquifer formation, the Shanxi formation sandstone fractured aquifer formation and the Quaternary loose pore aquifer formation. The thickness of the recoverable coal seam in the Late Carboniferous Taiyuan Formation is generally more than 10 m, and the maximum is approximately 30 m. The total thickness of the coal seam in the northern part of the Early Permian is 3–30 m, in the middle part is 2–10 m, and in the southern part is 0–7 m. In the late of early Permian, north China coalfields only had coal line or carbonaceous mudstone, the southern part had 13–16 coal layers, most of them can be mined, the total thickness of mining is approximately 18 m. In the Early Permian, from north to south coal seam gradually thickening, to 18–21 coal layers in Huainan coalfield, the total thickness reached 13 m. One of the most important coal-bearing strata in the study area is Taiyuan Formation, which is the nearest one besides the largest Middle Ordovician karst aquifers in this area. There is only 20–60 m Benxi formation between them. The mutual contact and fusion of Multi-aquifer, multi-coal seam and multi-aquiclude in the cross-section, formed a special stereo water-filled structure of coal field in north China.

The areas whose aquiclude thickness between bottom main coal seam and Middle Ordovician limestone top is less than 50 m are the thinnest in Beijing, Xishan, Jining, west of Xuzhou, eastern and southern foot of Taihang Shandong, and Huoxian, Hancheng and other ore districts. Calculate based on the vertical distance of the main mining coal seam from the top of the Middle Ordovician limestone, the ore districts whose floor aquiclude's thickness is less than 20 m are mainly Fengfeng, Huoxian and Hancheng ore district. According to the analysis of 1207 floor water-inrush incidents of the north China coalfields, the water inrush in the Taihang Shandong and southern foothills is the most serious, which is total 870 times, accounting for 71.4%. Secondry is the Xuhuai ore district in middle part of Shandong, which had 295 floor water-inrushes, accounting for 17.5%; the middle and southeastern part of Shanxi have the least number of water bursts occurred, only have 19 times, accounting for 1.5%. The distribution of the maximum amount of water inrush is consistent with the distribution of water inrush frequency, and it is mainly distributed in Taihang Shandong, southern foothill and even deep into the ore districts in middle part of Shandong.

2.3.3 Karst Development Characteristics

The regularity of karst development is the long-term research problem of many experts and scholars at home and abroad, they have delivered and published many important achievements. However, the development of karst is the result of the combined effect of various factors in the geological history period. The various factors have different status and roles in the different periods or different stages of karst development, so that the karst developments are different, they have similarities and uniqueness. The research object often has both general rule of existence and some special performance. Study the karst development law of karst aquifers which plays a role in coal water filling must combine the actual need of coal mining and fully consider the characteristics of karst aquifer in water filling the coal mine. When research on the syntagmatic relations of karst aquiclude, aquiclude and coal seam, should have a unified consideration on the distribution of karst aquifer's water abundance and water conductivity and recharge conditions and water filling characteristic. So, let the regular study of karst development directly serve the coal mining.

2.3.3.1 Relationship Between Karst Development and Lithology

The basic condition of karst development is the solubility. Under normal conditions, the higher the content of insoluble components such as clay minerals in the soluble rock, the lower the degree of development of dissolution of the rock. The calcite is usually easier to be dissolved than the dolomite, and the rock composed of coarse-grained mineral is more conducive to the occurrence and development of dissolution. The solution development degree of different karsts has different forms of expression, and the soluble rocks are mainly carbonate rocks, and gypsum layers may be sandwiched at certain horizons [16]. These gypsum layers have played or are playing a role that cannot be ignored in the karst development of carbonate rocks in adjacent horizons. The result of gypsum horizon's corrosion is the broken and caving-in of upper stratum. They strengthen the dissolution of carbonate, and are the important reason of the formation of brecciated carbonate rocks and limestone dolomitization. They also play an important role in special karst phenomenon like the formation of karst collapse columns in northern China.

2.3.3.2 Relationship Between Karst Development and Geological Structure

The tectonic action destroys the integrity of the soluble rock, and the karst development and water abundance near the fault zone are significantly increased. The fold with extension fracture and loose structural plane is conducive to karst development due to the crush of two sides; compressive torsion fault leads to compact structural plane which is not conducive to the migration of groundwater, and karstification is weakened. The rupture and its surroundings are often strong runoff of groundwater, and is a good area of karstification's concentrated development. The development of karst further promotes the circulation of groundwater, improves its movement condition and promotes more intense karstification, it is the dominant factor in the formation of karst development heterogeneity. The fault structure is not only the main channel of mine water inrush, but also in the process of water filling in the mine, the karstification along the fault zone can be re-intensified, and the deposits in some fault zones that have been clogged by loose deposits may be taken away by water so that the water conductivity can be improved. The deep groundwater in the stagnant state may communicate with the shallow water, so that the shallow water through the fracture in the deep supply the mine. These all will enhance the fracture's status in the development of karst and the role of water filling in the mine [17].

In addition, the existence of limestone, the presence of atmospheric precipitation, the presence of aggressive water, the distance from the recharge area of groundwater to the excrement area, and the runoff length of karst groundwater, the contact zone between the soluble rock and the insoluble rock are also important factors affecting karst development.

2.3.4 *Natural Water-Resisting Feature of Paleo-Weathered Crust on Unconformity Surface*

After the formation of the Cambrian Ordovician limestone deposits, the north China platform is lifted to land because of the Caledonian movement. Its top limestone generally suffered nearly 100 million years of weathering, erosion and dissolution, the holes cracks and other space of the limestone caused by erosion then filled by the later clay and sub-clay, forming an ancient weathering crust with natural water barrier properties. The water barrier property of ancient weathering crust on the top of Ordovician contains two meanings: First, the zone of krast's nondevelopment on the top of the Ordovician karst has a certain degree of confining water; the second is after the Kalidong movement the crustal rose, Ordovician limestone although developed the fissure after being weathered and eroded, but after filling, compaction and cementation, the filling section formed at the top of it is poor in water abundance, has high strength in mechanical properties, and it shows low permeability in groundwater flow. The relative filled aquiclude can also play a role of confining bed of water.

2.4 Water Control Characteristics of Structure in Coalfields of North China

The direct water-filled aquifer in the Carboniferous coal seam is mainly composed of the thin-bedded limestone aquifer formation or the middle-thick sandstone fractured aquifer formation. The two water-filled aquifers have a limited thickness and are often cut into small-scale fault blocks, or have thicker aquifer but worse permeability. The indirect water-filled aquifer in the Carboniferous Permian coal seams is a large-thickness karst aquifer in the Ordovician. Between these two layers it deposits a set of water-resisting stratum mainly composed of sandy shale, fine sandstone, shale and mudstone-based aquifer. The thickness of the aquiclude is different from each other in various ore district. Approximately in 82% ore district the thickness of the aquiclude is more than 20 m and in 50% ore district the thickness of the aquiclude is more than 50 m. Under normal circumstances, this aquiclude can resistant to the water pressure of the underlying karst aquifer.

Analyze in accordance to the above conditions, due to the limited water-bearing space and worse permeability of water-filled aquifer formation, and a thick aquiclude is deposited with the indirect water-filled aquifer. It is conceivable that these closed aquifer systems that not related to other water-filling aquifers are impossible to do have threats on the exploitation of Carboniferous Permian coal seams. However, the water-inrush accident in the north China coal field is frequent. The above assumption cannot be established, which means the direct water-filled aquifer is not an independent aqueous system, but an open or semi-open aqueous system that has closely hydraulic connection with the underlying indirect water-filled aquifer. And the water conduits are a variety of structural boundaries that control the close hydraulic connection of aquifer system. Therefore, the study of the occurrence regularity of karst water in north China coalfields under various geological structure conditions can provide a strong basis for evaluating and controlling mine water-inrush and mine water's resource utilization.

Carbonates are brittle rocks. Under different geologic tectonic conditions, apart from diagenetic fissures, structural fractures of different properties can be produced. By the dissolution of groundwater, they gradually form various rock fissure formation that has causal connection with geological structure. The size of the rock fissure, the degree of opening, the degree of connectivity, the degree of intensity and its changes in horizontal and vertical directions are all have differences, and the reservoir and movement laws of karst water are controlled. This geological structure that control the enrichment and distribution of karst water is known as karst structure water control.

2.4.1 Water Control Characteristics of Regional Structures

As early as the Proterozoic era, the region of north China has been consolidated into land platform. In early Paleozoic, it extensively deposited Cambrian, Ordovician carbonate series. The Caledonian movement has risen the area into land, and make the Cambrian and Ordovician limestone, which is widely distributed in this area, generally subjected to weathering. Until the Carboniferous, this area began to have sea baptism followed by slow retreat, in the north China area widely deposited land-sea interaction Carboniferous Permian coal-bearing strata. Due to the above reasons, the sedimentary cover of the north China platform is vast, and the sedimentary stability is stable, thus it forms various aquifer formation and the coal-bearing rock series in this area, which are widely distributed and stable. But from the Mesozoic, due to the Pacific Ocean's subduction to the Chinese mainland, the rigid, strong and stable north China land platform produced a series of north-east-oriented fracture and fault block up-and-down movements, so that the wholeness suffers damage. So that the huge complete Paleozoic aquifers that widely distribute in the north China area are cut into a series of large or medium-sized water structure.

The regional fault structure in north China, from west to east, are Guyuan fault zone, Lishi fault, Taihang mountain fault, Liaolan large fault, Tanlu large fault; from south to north, there are Jiyuan—Dangshan fault, Baotou—Jinzhou fault zone (Fig. 2.5). The above fault with large drop is large fault with water conductivity, it cuts the karst aquifers with the "zoning by across warp, zoning by warp" and controls the range of regional seepage field. The "zoning by across warp, zoning by warp" of regional fault divide the north China fault block into eight sub-level structural water storage unit: Helan—Liupan platform fold belt water storage unit, Ordos Taiao water unit, Shanxi fault-uplift water storage unit, Yanshan platform folded water storage unit, Liaodong platform water storage unit, north China fault depository water storage unit, Luxi fault-uplift water storage unit and Yuhuaitai water storage unit.

2.4.2 Water Control Characteristics of Ore District Structure

The karst water system and the water abundant area in the seepage field of the mine are controlled by the secondary geological structure and the derive fracture constitute the water control structure in the mining area. The control structure of the mine is often in the anticlinal core, covering the syncline wings and the turning points of the strata in two ends, the fault plate as well as the horst structure and the fault-intensive zone, and the development of solution crack to form a good water-abundance section. According to the characteristics of water storage and water control in the geological structure, the following types of water control are divided: monoclinal structure water control type, fold structure water control type, fault structure water control type and tectonic association water control type.

[1] Monoclinic water control type. Including monoclinic strata and fold wings, stratigraphic fissures and stratum fractures development, which are groundwater runoff channels. In the process of migration, the karst water influenced by the hydraulic gradient between the recharge area and the excretion area, and along the channel flows to the direction of the discharge zone. With the long-term effect in the bedding and karst fissure it gradually forms solution fissure. The karst groundwater is rich in dissolved pore and dissolved hole of monoclinic strata. In northern China, the typical monoclinic stratum water control structure has Shanxi Liulinquan karst system, the typical fold wing type has Hebei Kaiping syncline karst water system.

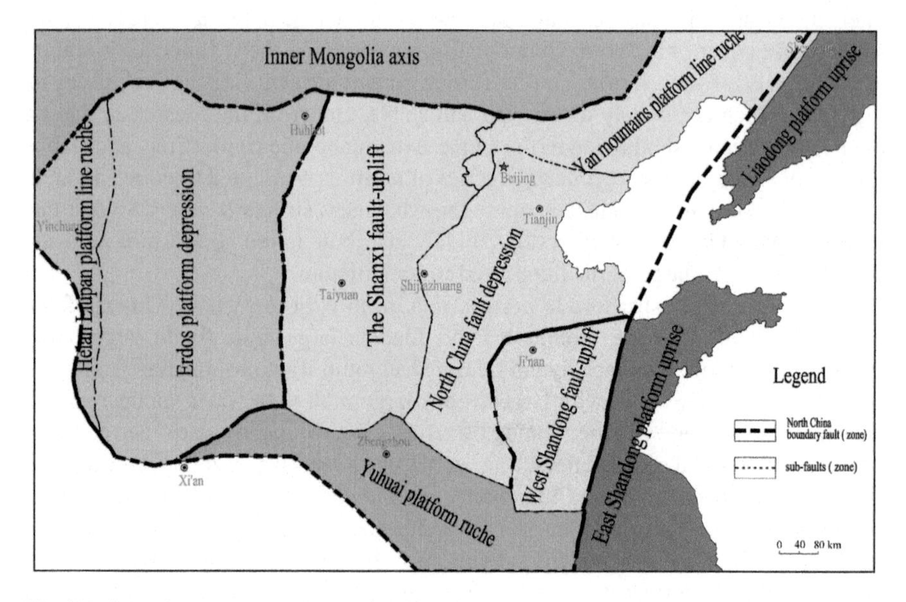

Fig. 2.5 Secondary tectonic units in north China

[2] Fold structure water control type. The fold structure includes the anticline, the syncline and the complex, the syncline structure, its pivot is horizontal or slant, it also can be crooked or undulant. Due to the brittle carbonate rocks, the pivot should concentrate in the joints, construct fissure development, in the turning point of fold, on the vertical rock layer it produces intense open fissure that parallel with axis; it intersects with the axis to become the two sets of shear fissures with 450° angle; and shear fissure developed into a vertical axial trace of the open fracture. Shear fissure surface is relatively smooth, the fissure is closed, with poor water permeability, but with the effects of multiple times being forced it can also be open. Therefore, in the rock bending parts, a variety of cracks interlinked which constitute a good reservoir structure and karst water channel. The north China fold structure water control includes: anticline fold water control type, the syncline fold water control, fold structure water control. Both Hebei Xingtai Baiquan anticline structure and Dahuoquan anticline water control structures are the typical fold structure water control type in this area.

[3] Fault structure water control type. Different regional tectonic faults have an important effect on the distribution and movement of karst water. There are a variety of regional structures in the regional tectonics in northern China, including crisscrossed fault assemblage, horst, and alternated ground mat clusters. The above-mentioned structure, whether it is water-resistant fault or water-conduit fault, are all developed with various properties of the fissures, and form the fault fracture zone, which is conducive to the occurrence or groundwater. Due to the different types and characteristics of fracture, water-abundance also has differences. The types of fracture structure water control are mainly divided into horst alternated with ground met cluster type water control and multi-group intensive fault structure water control.

2.4.3 Water Control Characteristics in Coal Mine

In recent decades, due to the lack of knowledge of the spatial distribution of the structure and the characteristics of the hydraulic properties, the hundreds of large and extra-large mine waterlogging accidents occurred in the various karst water mines in the north China coalfields. According to the spatial geometric characteristics of the tectonic distribution in the field, it can be divided into following three types.

2.4.3.1 Karst Collapse Column Type

In the lower part of the coal-bearing strata in the north China coalfields, there is a thick soluble rock generally, due to the long-term groundwater karstification, a large amount of karst cavities formed inside the soluble rock. Thus, the overlying strata becomes instable, and collapse to the dissolution of space, the cylinder formed because of collapse due to its profile like a column, it is called karst type collapse column. Whether the karst collapse column can conduct water, is mainly controlled by whether the voids in the karst collapse column are compacted or not. If the karst collapse column has great water-abundance and water conductivity, which leads to the relatively close hydraulic connection of different aquifers, and then brings great risk of water-inrush (Fig. 2.6).

Such as the water-inrush accident in the No.9 karst collapse column in 2171 working face in Kailuan Fangezhuang coal mine, the total volume of water inrush

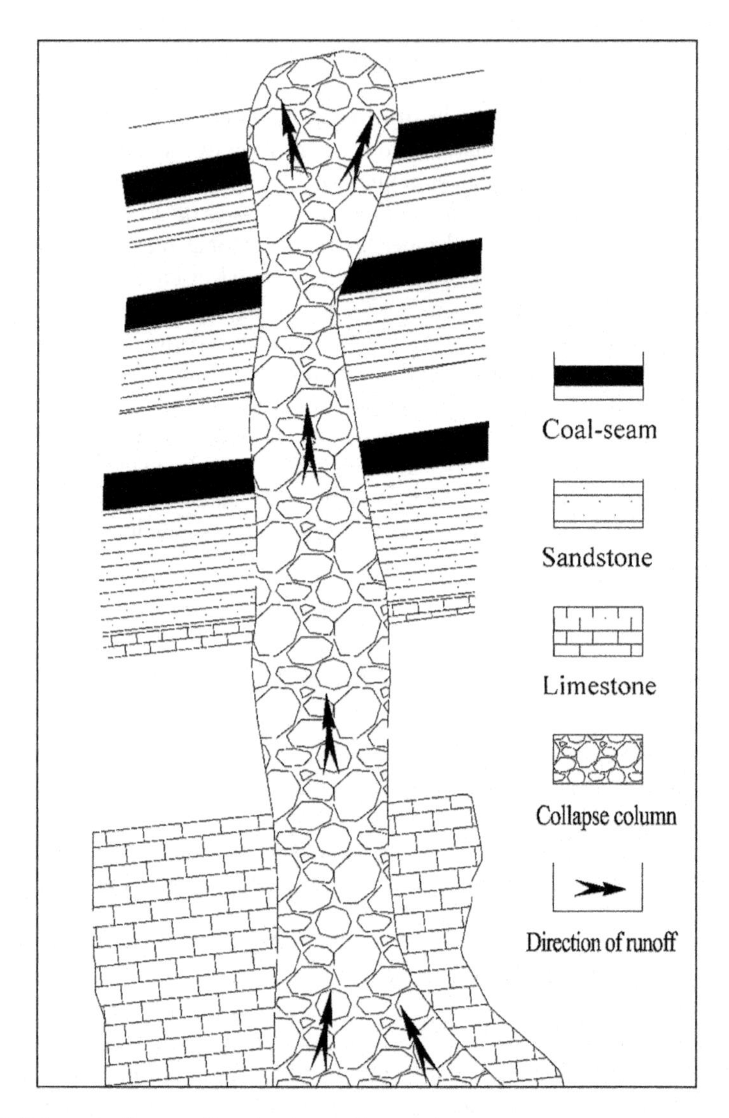

Fig. 2.6 Schematic diagram of a point-karst collapse column

in the peak of water inrush is 2053 m^3/min, which lasted 21 h, and submerged a mechanized mine with an annual output of 3.1 million t.

2.4.3.2 Fracture Type

The crustal stress near the faulted tectonic belt is concentrated, and the rock is broken, which destroys the integrity of the stratum. The rock in the stratum of false integration or unconformity is more broken, the fissure is more developed. It is possible to form a gushing channel through a tectonic zone with a concentrated tectonic zone or a weak fake or unconformity. The hydraulic connection between the working face and the aquifer will lead to a huge safety hazard (Fig. 2.7).

By analyzing the sedimentary regularities and hydrogeological conditions of the north China Coalfield, the north China coalfield under the action of multi-stage crustal stress, its brittle aquiclude release crustal stress in the formation of breaking, which leads to that aquiclude creates asymmetrical intensive joints and fissures, and forms surface fissure network. With the increase of the head difference of the upper and lower aquifers which contains surface fissure network, the hydraulic connection between the two aquifers will gradually increase. This type of surface fissure network structure water control situation has been confirmed in Donghuantuo Mine of the Kaiping coalfield.

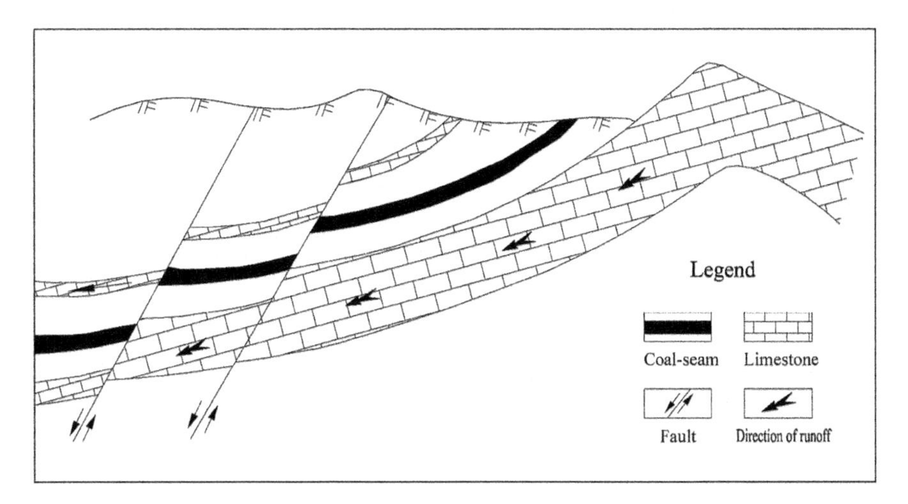

Fig. 2.7 Schematic diagram of linear fractures

2.4.3.3 Sub-outcrop Type

The water—filling aquifer and carbonate—water—filled aquifer in the coal—bearing stratum of the north China coalfield are always in unconformable contact with the Quaternary loose sedimentary stratum in narrow strip. Whether the hydraulic connection between the Quaternary loose pore aquifer and bedrock fissure water aquifer is close mainly depends on: (1) whether the concealed outcrop in the unconformity contact surface is strongly weathered; (2) Whether there is a thicker cohesive soil aquiclude between the Quaternary loose sedimentary stratum and Bedrock fissure water aquifer (Fig. 2.8).

2.5 Geological Structural Model of Water-Soluble System and Coal-Bearing Stratum

At different stages of coal mining, the disturbance of the mining activity to the groundwater system will be different. In the capital construction period which is mainly composed of excavation of coal and rock and extension engineering, the main disturbance is to destroy the direct water and water layer subsystem of coal seam. In the process of mining, the span and space of the working face are much larger than that of the excavated coal roadway. When the stress balance of the rock around the goaf is destroyed, which results to the bulge of the bottom plate, deformation and fracture

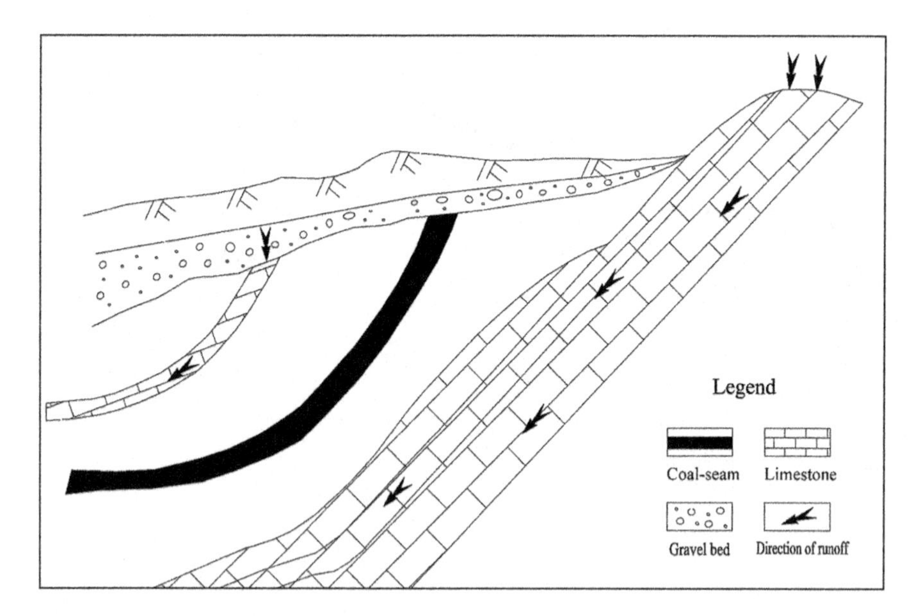

Fig. 2.8 Schematic diagram of sub-outcrops

of the roof. Professor Li Baiying and others have proposed a theory that coal seam floor damage "the under three belts", that is, in backstoping process the coal seam floor from top to bottom is divided into water-conductivity damage zone, effective aquicludes protection zone and pressurized water guide belt. The damage of the coal-bearing stratum can cause the change of the geological structure and the groundwater seepage. The aqueous system of the mining area mainly receives the supply of the fissure water conduction, and the runoff direction is mostly vertical movement. The drainage system of the aqueous system is mainly the artificial drainage of the mine. If thing go on like this, the groundwater system of ore district under the natural state gradually changed from horizontal to vertical movement. This phenomenon is "mine hydration" phenomenon, also known as "mine hydration" system [18].

The regional fault block structure of the north China platform makes the distribution of coal field strongly consistent with the distribution of karst water system. In this paper, the karst water system is used as a unit to study the problem of water inrush of the coal seam floor. With the control of the hydrogeological conditions of the coalfield, different karst water systems have different degree, mode and approach of the water inrush in coal mine floor. To reveal the general laws of the coal seam floor water inrush in the Karstic Permian of the karst water system under the control of hydrogeological conditions, the following classify and analyze from the karst karst water system and the geological structure of coal-bearing strata.

In the same structure mode, the occurrence of karst water aquifer and the characteristics of groundwater seepage are similar, thus the modes of the karst water's filling, running and discharging and have similar characteristics with the development characteristics of karst development. And the karst aquifer under each mode is similar with the geological structure of coal-bearing series, so it shows similarity to the impact of coal mining [19].

Based on the above-mentioned sedimentary regularity of the north China coalfield, the structural characteristics of the karst water system and the water control system analysis, and according to superposition relationship between the geological structure of the coal-bearing strata and the characteristics of the karst groundwater flow field, the geological structure models of karst water system of main ore district and coal-bearing stratum are divided into five structural types: monoclinic order type, monoclinic inversion, parallel type, syncline-basin type and fault-block and other five structural modes. The corresponding relationship between main coal field and karst water model in north China is shown in Table 2.2.

Table 2.2 The relationship between coalfield and Karst water model in north China

Coalfield	Karst water system	System model	Coalfield	Karst water system	System model
Hanxing coalfield	Baiquan and Heilongdong spring area karst water system	monoclinic order type	Huaibei coalfield	Huaibei karst water system	Fault block and other types
Hedong coalfield	Tianqiao, Liulinquan karst water system	monoclinic order type	Jingxing coalfield	Weizhou spring area karst water system	Monoclinic inversion type
	Yumenkou spring area karst water system	monoclinic inversion type			
Xuzhou coalfield	Xuzhou karst water system	Fault block and other types	Feicheng coalfield	Feicheng karst water system	Syncline-basin type
Taozao coalfield	Shiliquan area karst water system	Syncline-basin type	Pingdingshan coalfield	Pingdingshan karst water system	Syncline-basin type
Zibo coalfield	Yangshuiquan area karst water system	Monoclinic order type	Kailuan coalfield	Tangshan karst water system	Syncline-basin type
Hancheng coalfield	Hancheng karst water system	Directional type	Anyang, Hebi coalfield	Xiao nanhai, pearl spring, Xujiagou spring area karst water system	Monoclinic order type
Shanmian coalfield	Mianchi spring area karst water system	Directional type	Jingxi coalfield	Yuquanshan spring area karst water system	Monoclinic order type
Wuhaikabuqi coalfield	Zhuozishan spring area	Directional type	JIncheng coalfield	Sangu spring area karst water system	Monoclinic inversion type
JIdong coalfield	Jinan, Mingshui spring area karst water system	Monoclinic order type	Yanlong coalfield	Yanlong karst water system	Monoclinic order type

(continued)

Table 2.2 (continued)

Coalfield	Karst water system	System model	Coalfield	Karst water system	System model
Tongchuan, Chengou coalfield	Yuanjiapo, Wentang, Fenquan spring area karst water system	Monoclinic inversion type	Taiyuan Xishan coalfield	Jinsi spring area karst water system	Monoclinic order type
Jiaozuo coalfield	Jiulishan spring area karst water system	Monoclinic order type	Yangquan coalfield	Niangziguan spring area karst water system	Monoclinic inversion type
Qinshuiyuan coalfield	Yanhe spring area karst water system	Monoclinic inversion type	Lingshan coal mine	Shuimocao spring area karst water system	Syncline-basin type
Xingyang coalfield	Xingyang karst water system	Monoclinic order type	Mixian coalfield	Chaohua spring area karst water system	Syncline-basin type
Huoxi coalfield	Guozhuang spring area karst water system	Monoclinic order type	Ningwu coalfield	Majuan Leimingsi spring area karst water system	Monoclinic inversion type

2.5.1 *Monoclinic Order Type of Structural Patterns and Features*

The characteristics of the monoclinic order type structure are: the submerged karst aquifer and the coal seam stratum's angle are all placid and have a monoclinic structure. The flow direction of the aquifer in the floor is generally the same as that of the coal seam stratum. Its flow field characteristics are mostly in the formation of fan-shaped and flow to the main drainage point. The coal-bearing stratum is generally distributed in the downstream of the aqueous system (Fig. 2.9). The most important recharge of karst water system comes from infiltration of atmospheric precipitation. Typical "monoclinic order type" structure model forming overflow springs for the confining bed, karst water discharge is more concentrated. The karst aquifer is enriched in the interface between the Ordovician and the coal seam, and the large karst water can form a strong runoff belt at this contact surface. The complete monoclinic order type has a distribution of large-scale Carboniferous Permian coal-bearing strata downstream, and it is one of the riskiest types of coal mine water inrush.

The monoclinic structure model has the largest amount of distributing in the north China coal field, and accounts for 40% of the main coal field in the thirty main coal fields in north China. It is mainly distributed in the west side of Lvliang Mountain, Taihang Mountains, West Henan and middle and southern part of Shandong mining area. In the downstream of the monoclinic order type structure. There is a large area of Carboniferous Permian coal bearing strata confined area. The lower coal is generally located below the regional karst water level, and in the contact zone of coal-bearing strata and carbonate, karst water has strong water abundance. Hanxing coal field, Jiaozuo coalfield and Zibo coal field and other large water mines are all belong to this structural model. The strong water abundance and large water pressure of karst water lead to the fact that this model is the most threatening type for karst water inrush. It is generally characterized by large scale of water inrush, strong water potential and serious loss to mine. For example, May 13, 1935 Shandong Zibo Beida coal mine water disaster; June 4, 1960 Hebei Fengfeng coal mine water disaster; 1979 Henan Jiaozuo Yin Ma coal mine with the maximum waterfall of 14,580 m^3/h; 1981 Henan Hebi Macun coal mine with the amount of water inrush 13,500 m^3/h [19].

2.5.2 *Monoclinic Inversion Type of Structural Patterns and Features*

The characteristics of the monoclinic inversion type's structural patterns are summarized as follows: the inclination of the coal-bearing strata and the karst aquifer is relatively placid, and it belongs to the monoclinic structure. The flow direction of

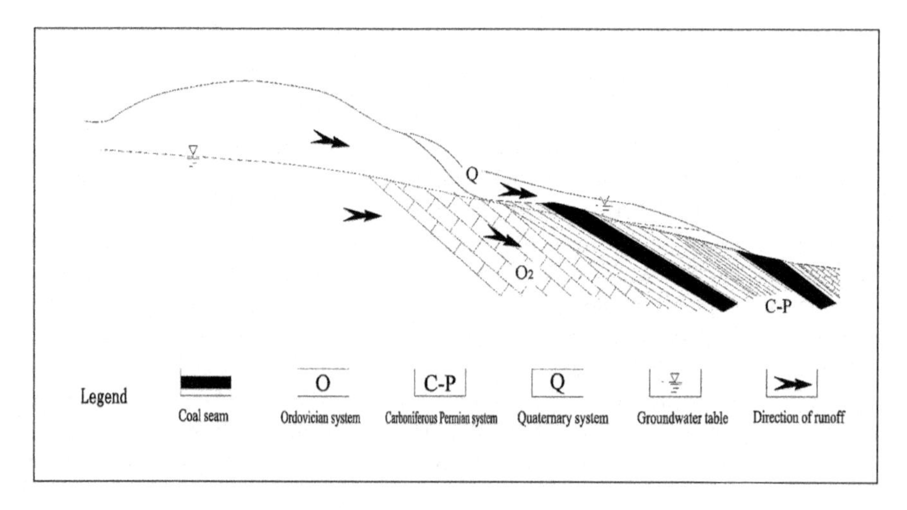

Fig. 2.9 Schematic diagram of monoclinic order type structure mode

the aquifer in the floor is opposite to that of the coal-bearing stratum. Its flow field morphology is always in form of fan-shaped to concentrate to the main discharge point. The complete monoclinic inversion system model coal-bearing stratum is in upstream of the system (Fig. 2.10).

The monoclinic inversion structure is widely distributed, it distributed on the west side of the Taihang Mountains and the northwest side of Fenwei graben. The typical monoclinic inversion type coal field has Jincheng coal field, Hedong coal field and Yangquan coalfield, etc. In the thirty main coal fields in north China, the coalfield of this structural model account for 27%, and in the five structural modes, it is only second to monoclinic order type. Although like the monoclinic order type, the karst water is also enriched in the contact zone between carbonate rocks and coal-bearing strata, due to the high-water pressure area is often inconsistent with the karst development area, this model has less threats on water inrush. Only when the tectonic development, it will form a water inrush event. For example, there are 30 large faults in the Xuangang mining area, especially in the east-north of the spring area, along the Sanggan River and the Hutuo River in north-to-east watershed parallelly develops a series of faults, whose two sides are uneven to north or to south, and the middle is arranged with the basement, graben, which cut the ground mat into the north-east of the strip-shaped rock. The Jiaojiazhai Mine in the system once occurred the water burst accident, among which Liu Jiuliang Mine in 1973–1988 years had a total of 16 flooding accidents in 15 years, in May 1986 the maximum amount of water up to 8366 m^3/d. On August 7, 2011, the Sangshuping Coal Mine of Shaanxi Hancheng mining company that located in Yuanjiapuquan Yiwen Tangquan karst water system occurred flooding accident, resulting in flooding the entire mine. In the process of excavation of Dongshan coal mine in Niangziguan spring area, 332 faults were found, and the number of faults that more than 5 m are 51, among which mostly are normal faults. This mine occurred water inrush accident in 1985. The initial volume of water inrush was nearly 13,000 m^3/d [20, 21].

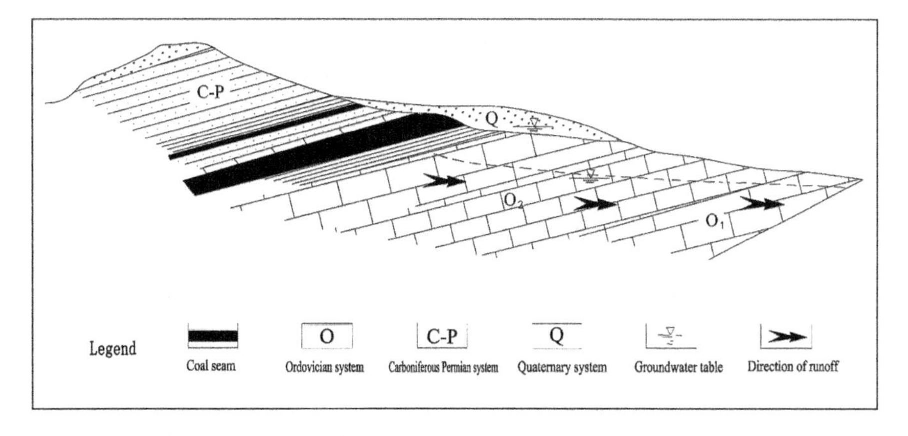

Fig. 2.10 Schematic diagram of monoclinic inverse type structure mode

2.5.3 Syncline Basin Type of Structural Patterns and Features

The syncline basin type structure is characterized by the fact that the system consists of a relatively complete syncline structure or a small faulted basin. From the plane point of view, the structure's groundwater flow field and coal-bearing strata are concentric. Groundwater and surface water, have shown a confluence towards the center of the basin (Fig. 2.11). The karst water aquifer drainage point has the characteristics of scattered distribution, mostly excreted along the areas which is abundant with rivers of synclinal axis or center of basin. There is a more complex relationship of recharge and discharge between Karst water, pore water and surface water.

Mainly distributed in the Taihang Mountains, Yanshan border area, Yanshan piedmont and other mining areas, in the mountains of middle part of Shandong there are a small amount of distribution. For example, Kailuan coal field, Pingdingshan coal field, Feicheng coal field. There are lateral, vertical, direct and indirect and many forms of transformation between karst water in the inside edge of basin and coal-bearing aquifers in the basin. The highest-pressure zone in coal is often also karst groundwater enrichment area [22]. Therefore, water inrush accidents occur frequently during coal mining, and severe water permeation accidents often occur under this karst hydrogeological condition. For example, in 1984, the water inrush accident occurred in Fangezhuang coal mine of Tangshan Kaiping syncline karst water system, whose maximum amount of water inflow up to 123,180 m^3/h, with a direct loss of 560 million yuan; the coal mine water in Feicheng basin karst water system had water inrush accidents for many times: In 1993, the water inrush in Guozhuang coal mine, with the maximum water inflow of 32,970 m^3/h, in 1997 Longzhuang Coal Mine water inrush occurred, with the maximum water inflow 15,000 m^3/h; in 1985, the water inrush in Taoyang coal mine water with the maximum water inflow 1794 m^3/h. In addition, the Yufengshan coal mine in karst water system of

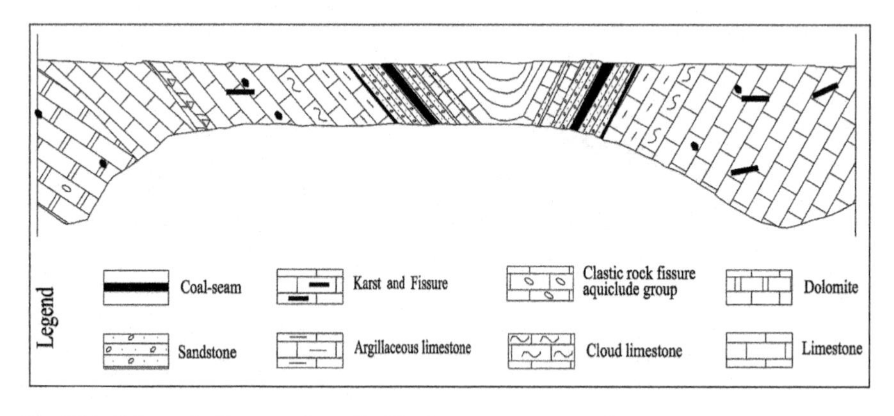

Fig. 2.11 Schematic diagram of synclinal basin type structure mode

Shuishentang-Qilihe spring area occurred water inrush in 1977, with the maximum water inflow 600 m^3/h.

2.5.4 Directional Type of Structural Patterns and Features

The directional type structural pattern is characterized by the fact that the dip angle of karst aquifer and coal—bearing strata in this model is steep and is monoclinic structure. The karst groundwater flow field is mostly rectangular, and the most characteristic of the directional system pattern is that the direction of coal-bearing strata is consistent with the direction of karst groundwater main runoff (Fig. 2.12). The groundwater of this pattern is mainly subject to atmospheric precipitation[c]. Due to the small amount of carbonate exposed in the system, the recharge from atmospheric precipitation to groundwater is limited. The discharge points of directional structure are relatively concentrated, mostly are in the slanting side of the aquifer, the form of discharging is spring excretion.

The directional pattern often appears in areas with strong tectonic deformation, such as the north and south ancient ridge belt area in the west edge of the Ordos basin, the Yanshan area and the Huainan area. Such as the Shaan Mian coalfield in the Mianchi spring area karst water system, the Wuhai Kabu coal field in the Zhuozishan spring area, the Hancheng coal field in the Hancheng karst water system. The structure of the directional type is similar to that of the monoclinic order type. Its influence on karst water is mainly manifested in the water quantity, but it is controlled by the smaller scale of karst water system, the water volume of mine is relatively small. However, when the depth of coal mining is great, the sudden influx of gray limestone water of the thick layer will cause serious accidents. For example, on March 1, 2010 the water inrush accident occurred in Inner Mongolia Wuhai camel

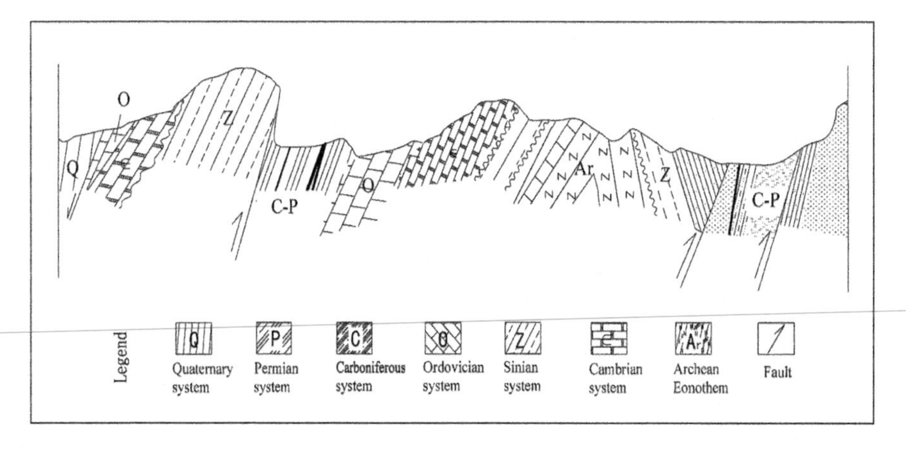

Fig. 2.12 Schematic diagram of sync strike-type structure mode

mountain mine, in which the maximum amount of water 79,000 m^3/h, and resulted in 31 people being killed in the accident.

2.5.5 *Fault Block and Other Types of Structural Patterns and Features*

The karst aquifers and coal—bearing strata of the fault block and other types are shown in block fault, and the tectonic development and karst groundwater discharge are scattered. The recharge relationship between karst water and coal seam formation of this system is complicated, and coal mining has the influence of water quantity and water quality on karst water. However, due to the relatively small size of the system, the relatively low degree of concentration of karst groundwater, therefore, there is a certain risk of water inrush, but the large water inrush (water inflow > 200 m^3/min)

Table 2.3 The summary table of karst water system of coal seam floor in north China

Project	Pattern				
	Monocline order type	Monocline inversion type	Directional type	Syncline-basin type	Fault block and other types
Relationship between coal—bearing strata tendency and karst water flow	Consistent	Reverse	Parallel direction	Converge to the middle of the basin or to the oblique axis	Unsure
Karst aquifer formation	Placid monoclinal	Placid monoclinal	Steep and declined monocline	Rifted basins or synclinal	Block fault
Main recharge source	Precipitation	Precipitation, River	Precipitation	Precipitation	Precipitation
Groundwater flow pattern	Fan-shaped	An-shaped	Rectangle	Concentric circles	Unsure
Hydrodynamic distribution	Recharge—runoff—excretion—pressure stagnation	Recharge—runoff—confluence—excretion	Recharge—diameter sparse—excretion—pressure stagnation	Recharge—pressure confined—excretion	
Distribution orientation of coal—bearing strata	Downstream of system	Upstream of the system	Side of system	Middle of the system	Unsure
Coal mine water inrush		General	General	Serious	

is rare. The coal mines of typical fault block and other type system in north China are Xuzhou coal field in Xuzhou karst water system, Huaibei coal field in Huaibei karst water system.

The characteristics of different structural modes of coal-bearing strata and karst water system in north China coalfields are shown in Table 2.3.

2.6 Summary

In this chapter, the hydrothermal geology characteristics of the north China coal field are analyzed based on the sedimentary regularity of the coal-bearing strata, the characteristics of the karst water-filling system and the tectonic water control. Based on the overlapping relationship between the geological structure of coal—bearing strata and the flow field characteristics of coal—filled strata, the geological structure model of karst water system and coal-bearing strata in the main coal ore district in north China is classified into five types, which laid the foundation for the establishment of the hydrogeological conceptual model of the risk of water inrush from coal floor.

References

1. Liang, Yongping, Xingrui Han, and Jian Shi. 2005. The Karst Groundwater system in the peripheral area of Ordos Basin: Its patterns and characteristics. *Actageoscientica Sinica* 26 (4): 365–369 (in Chinese).
2. Liang, Yongping, and Xingrui Han. 2013. *Environmental problems and protection of karst groundwater in northern China*. Beijing: Geological Publishing House (in Chinese).
3. Liao, Zisheng. 1978. *The main characteristics of karst in North China and the main types of karst water storage structure*. Beijing: Geological Publishing House (in Chinese).
4. Liu, Guangya. 1979. *Bedrock groundwater*. Beijing: Geological Publishing House (in Chinese).
5. Liu, Qiren, Fengqi Zhang, and Yifang Qin. 1989. *Evaluation and prediction of karst groundwater resources in northern China*. Beijing: Ministry of Geology and mineral resources (in Chinese).
6. Lu, Yaoru, Fenge Zhang, and Changli Liu et al. 2006. Karst Water resources in typical areas of China and their Eco_hydrological characteristics. *Actageoscientica Sinica* 27 (5): 393–402 (in Chinese).
7. Zhang, Yunjun. 2012. *Study on comprehensive water separation performance and water inrush risk prediction of coal seam floor*. Beijing: General Research Institute of coal science (in Chinese).
8. Shao, Aijun, Jianping Peng, and Zhiguang Li et al. 2011. *Floor water inrush in coal mine*. Beijing: Geological Publishing House (in Chinese).
9. Hai-ling, Komg, Xie-xing Miao, and Lu-zhen Wang. 2007. Analysis of the harmfulness of water-inrush from coal seam floor based on seepage instability Theory. *Journal of China University of Mining and Technology* 17 (4): 453–458.
10. Li, Jinkai. 1990. *Prevention and control of mine karst water*. Beijing: China Coal Industry Publishing House (in Chinese).
11. Wang, Xiulan, and Zhongxiu Liu. 2007. *Mine hydrogeology*. Beijing: China Coal Industry Publishing House (in Chinese).

12. Pan, Yuanbo. 1986. Derivation of the safety head formula for water masses. *Journal of Hefei University of Technology* (*Natural Science*) 8 (1): 99–103 (in Chinese).
13. Qiang, Wu, and Entai Guan. 2009. Emergency responses to water disasters in coalmine, China. *Environmental Geology* 58 (1): 95–100.
14. Mironenko, V., and F. Strelsky. 1993. Hydrogeomechanical problems in mining. *Mine Water and the Environment* 12 (1): 35–40.
15. Zhang, Xiangdong, Han Dawei, and Liu Shijun. 1997. Mechanism of water inrush from coal seam floor and distribution characteristics of "lower three zones". In *Conference on New Development of Mine Construction and Geotechnical Engineering* (in Chinese).
16. Shen, Guanghan, Baiying Li, and Ge Wu. 2009. *Theory and practice of special mining in coal mine*, 56–72. Beijing: China Coal Industry Publishing House (in Chinese).
17. Wang, Chengxu, and Hongmei Wang. 2004. Thinking on theory and practice of coal mine water control. *Coal Geology and Exploration* 32 (Z1): 100–103 (in Chinese).
18. Gao, Yanfa, and Shi Longqing. 1999. *Water inrush law of floor and water bursting dominant surface*, 48–68. Xuzhou: China University of Mining and Technology press (in Chinese).
19. Wu, Q., Y. Liu, and L. Luo. 2015. Quantitative evaluation and prediction of water inrushvulnerability from aquifers overlying coal seams in Donghuantuo Coal Mine, China. *Environmental Earth Sciences* 74: 1429–1437.
20. Wu, Q., and S. Ye. 2008. The prediction of size-limited Structures in a coalmine using artificial neural networks. *International Journal of Rock Mechanics and Ming Sciences* 45 (6): 999–1006.
21. Wu Q, S. Xie, Z. Pei, J. Ma. 2007. The new practical evaluation method of floor water inrush. III: application of ANN type vulnerability index method based on GIS. *J Chin Coal Soc* 32 (12): 1301–1306.
22. Wu Q, J. Wang, D. Liu, F. Cui, and S. Liu. 2007. The new practical evaluation method of floor water inrush. IV: application of AHP type vulnerability index method based on GIS. *J Chin Coal Soc* 34 (2): 233–238.

Chapter 3
Acquisition and Quantification of Main Controlling Factors of Water Inrush from Coal Seam Floor in Coalfields of North China

The water inrush from coal seam floor is a dynamic phenomenon affected by the combined action of hydrogeology and mining with complex nonlinear dynamic characteristics under the control of multiple factors, resulting from comprehensive function of geologic field, stress field and seepage field. The mining of coal seams destroys the balance of these three fields, thus causing them to change and rebalance. It will lead to water inrush accident once conditions, water bursting source, water passage and water inrush intensity, mature. To systematically study the complex mechanism of the water inrush from the coal seam floor and reveal the complex nonlinear dynamic process of floor water inrush, it is necessary to comprehensively analyze the geological field, stress field and seepage field. To grasp the main contradiction of the problem, it is necessary to analyze the main controlling factors of the water inrush its index system, and to construct the conceptual model of hydrogeology of water inrush from coal seam floor, laying the foundation for the research and development of the new evaluation model which conforms to the actual coal seam floor water inrush.

3.1 Main Controlling Factors of Water Inrush from Coal Seam Floor

The water pressure of the confined aquifer is the source of the water inrush from coal seam floor and the direct cause of forming the water passage. The water abundance of the confined aquifer is the base of water inrush, which directly determines the water inrush would occur or not and the bursting water quantity. Geological structure (faults, folds and collapse pillars, etc.) is often a good channel for groundwater, which can directly lead to the occurrence of water inrush. The aquifuge has played an impedance role, and its total thickness, lithology, lithological combination and other factors determine the impedance performance [1]. Mining activity and changes in mine pressure induce the occurrence of floor water inrush. So, in this paper, the

Y. Zeng, *Research on Risk Evaluation Methods of Groundwater Bursting from Aquifers Underlying Coal Seams and Applications to Coalfields of North China*, Springer Theses, https://doi.org/10.1007/978-3-319-79029-9_3

main control factors influencing the water inrush from coal seam floor in the north China coalfield are determined from the comprehensive analysis of the confined aquifer, the aquifuge, the geological structure, the mining works and the ancient weathering crust.

3.1.1 Confined Aquifer

The confined aquifer can not only provide the water source for water inrush from coal seam floor, but also supply the damage power to help to shape the water passage. The main influencing factors are water pressure, water abundance and permeability of the confined aquifer.

(1) **Water pressure**

Water pressure of the confined aquifer is the head pressure acting on the lower confining bed of the aquifuge, and the corresponding head height is the difference between the elevation of the free water interface and the elevation of top interface of the confined aquifer. The water pressure of the confined aquifer can damage the confining bed which resists water inrush, and it is power of the water inrush from coal seam floor. Therefore, in the normal area of the direct water channel without structural fissure development, the water pressure must be large enough to be able to destroy the water resisting layers to cause water inrush from coal seam floor. In addition, the precondition of water inrush caused by coal mining is that the interface level of free water must be higher than the level of coal seam floor. The influence of the water pressure on the confining bed is mainly manifested through hydrostatic pressure and hydrodynamic pressure.

(2) **Water abundance**

Water inrush from coal seam floor requires the water content of the aquifer to meet certain limits. Factors affecting the water yield property are mainly aquifer thickness, unit water inflow, aquifer permeability and so on. The main aquifers in north China are mainly composed of Archaean limestone aquifer and Ordovician limestone aquifer. Often due to the limited thickness of the Archaean limestone aquifer and its general water abundance, it is not a threaten of water inrush from coal seam floor. However, the Ordovician limestone aquifer is a big threaten because of its large thickness of aquifer and better water abundance. If there is a channel conduction between these two aquifers, causing the Archaean limestone with bigger water pressure and greater water abundance, it is easy to lead to water inrush accident.

3.1.2 Aquifuge of Coal Seam Floor

Aquifuge of coal seam floor is the aquifuge between the coal seam floor and the top interface of the confined aquifer, playing an inhibitory effect on water inrush from coal seam floor. The main factors influencing the water barrier performance of the aquifer are: the total thickness, the lithologic combination, the key layer position and so on.

The main factors influencing the water barrier performance of the aquifer are: the total thickness, the lithologic combination, the key layer position and so on.

(1) **Total thickness**

The total thickness of the aquifer is an important parameter to measure the water barrier performance of the aquifer. The greater the total thickness of the aquifer, the stronger the water barrier performance. The traditional waterflooding coefficient method is used to evaluate the concept of hydrological and geologic physics when the coal seam is inrush. The main consideration is the consideration of the water pressure and the thickness of the seams in the coal seam. With the deepening of the coal seam in the north China coalfields, the thickness of the aquifer between the coal seam and the confined aquifer is getting smaller and smaller, and the possibility of water bursting in the coal seam is also increasing.

(2) **Lithology group**

Stratigraphic lithology is not the same as the corresponding mechanical strength, the ability of its resistance to water pressure is not the same. Thus, the same thickness of the case, by the different lithology of the formation of its water barrier performance is not the same. In general, the greater the mechanical strength of rock formation, the stronger the ability of water pressure; the other hand, the weak resistance of water pressure. When the rock is a plastic rock, its water barrier is relatively good; and the rock is brittle rock, its water barrier is relatively weak. The water barrier properties of the formation need to consider the mechanical strength and brittle plastic rock water barrier [2]. The lithology of the coal seam in the north China Carboniferous Permian coal seam is mainly composed of four lithologies: shale, sandstone, bauxite and iron rock, considering its mechanical properties and water barrier properties. From strong to weak: sandstone, iron rock, bauxite, shale. Natural water dikes are often composed of multi-layer formation, when the brittle rock and plastic rock layer, the water blocking effect is the best.

A great deal of research has been done on the water cut performance of different lithologic strata, which is divided into three cases:

[1] Quality equivalent. With the mudstone as the standard, the other lithologic water resistance can be converted into equivalent mudstone thickness according to the mass equivalent coefficient, and the equivalent thickness of each section is the equivalent thickness of the formation water resistance (Table 3.1).

Table 3.1 The equivalent coefficient of mass

Lithology	Mudstone	Limestone	Sandstone	Sandy shale
Equivalent coefficient of mass	1.0	1.3	0.8	0.4

Table 3.2 The equivalent coefficient of strength

Lithology	Sandstone	Limestone	Sandy shale	Mudstone	Crushed zone
Equivalent coefficient of strength	1.0	1.2	0.7	0.5	0.35

[2] Strength equivalent. With the sandstone as the standard, the other lithologic water resistance can be converted into the equivalent sandstone thickness according to the strength equivalent coefficient, and the equivalent thickness of each section is the equivalent thickness of the formation water resistance (Table 3.2).

[3] Comprehensive equivalent. Considering the quality and strength of the rock formation, the equivalent water quality of the lithologic strata is equivalent to the equivalent coefficient, and the equivalent thickness of the rock is obtained (Table 3.3).

(3) **Position of key stratum**

The key stratum is a stratum with the highest strength of the aquifuges. The key layer of different positions, the water blocking capacity is not the same. In general, after coal mining, due to geological stress and mine pressure, the coal seam floor will produce a certain depth of the mine pressure damage; further combined with the seepage field, in the confined water will produce a certain height of the aquifer; (Fig. 3.1a and c); only when the key layer is located in the barrier layer, the critical layer is located at a distance In the middle of the water layer, the upper and lower strata are plastic rock, the water blocking effect is the best, really play the role of the key layer of water (Fig. 3.1b).

Table 3.3 The comprehensive equivalent coefficient

Lithology	Sandstone	Limestone	Sandy shale	Mudstone	Crushed zone
Comprehensive equivalent coefficient	1.1	1.2	0.7	0.6	0.3

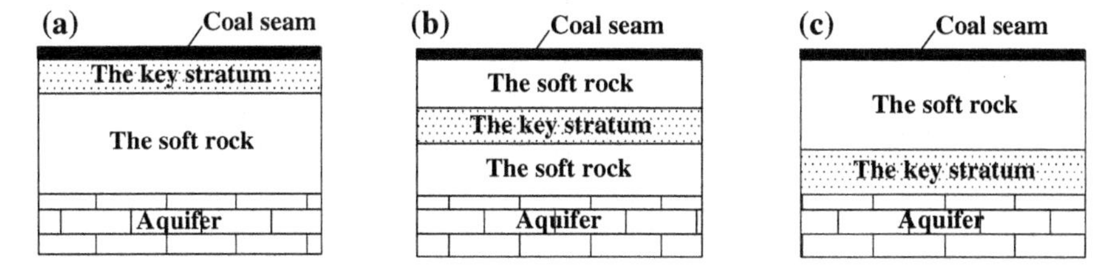

Fig. 3.1 Schematic diagram of the key strata location

3.1.3 Geological Structures

3.1.3.1 Faults

The effect of fault on floor water inrush is mainly manifested in the following:

[1] Good water guide channel: When the coal mining is close to or expose to the faults and the faults connect seam that with great water abundance, due to the broken rock in faults and stress's destroying effect caused by coal mining, the faults tend to form a good water conduct for the groundwater, which directly cause water inrush accidents: even if the fault not directly conduct water, it may lead to delayed water inrush accidents [3].

[2] The thickness of the aquiclude becomes smaller: The relative displacement of the two sets faults often shortens the distance between the two sets coal seams and the aquifer (Fig. 3.2a). When the drop of fault is large, the coal seam and the aquifer may be docked, and the water inrush accident is likely to occur (Fig. 3.2b); especially when the fault is connected multiple aquifers, it increases the risk of water inrush (Fig. 3.2c).

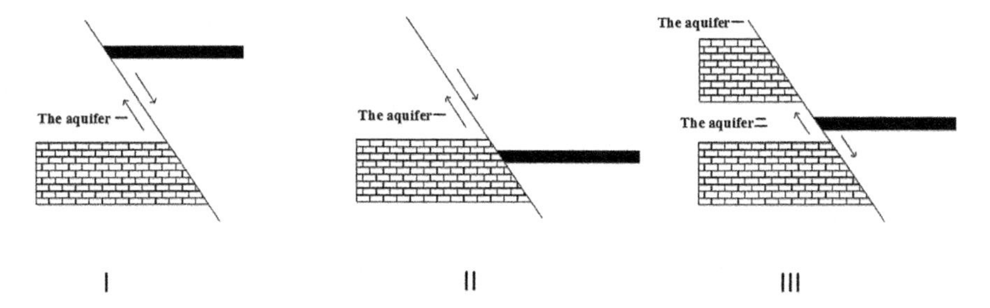

Fig. 3.2 Schematic diagram of contact relationship between seams and aquifers under different fall heads

[3] Lower water resisting intensity: Under the action of tectonic stress, the rock in fault zone broke and its mechanical intensity also decreased greatly, which results to great decrease of water-resisting property of the fault which is only approximately 30% of the normal lot, and it becomes the water inrush black spot.

[4] Increase the height of the raising and the depth of mine pressure damage: The fault is further damaged by the mine pressure damage caused by the coal mining. Due to its low mechanical intensity, the damage depth of mine pressure and the height that confined water raise all increase, which can easily lead to water inrush [4].

3.1.3.2 Collapse Column

There are many genetic mechanisms of the trapped column, the rock mass in the column is high, and the mud is cemented, and the long-term softening and erosion of the confined water lead to the greatly reduced water barrier performance, which is often a good channel for groundwater. The existence of the column makes the stress of the surrounding rock be released, resulting in the development of the rock fissure, coal mining will further aggravate the destruction of the surrounding rock, water permeability will be further enhanced. It is very easy to cause large water inrush accident in the Ordovician limestone aquifer, and it is necessary to have advanced concealment and take precautionary measures.

3.1.3.3 Fold

The formation of the fold shaft and its adjacent strata are more concentrated, the fissure is more developed, and the corresponding permeability is stronger. After the special coal seam is mined, the strata will be further damaged along the fold axis. When the foliage is connected with the aquifer, it is often a good channel for the aquifer, triggering water inrush accident, should strengthen the prevention.

Not all structures will lead to water inrush accidents, geological tectonic rupture of water needs two major factors: First, there are aqueducts; Second, the structure of aquifer need to have a strong water-rich [5].

3.1.4 Disturbance of Mining

The excavation works caused the destruction of the strata, changed the original stratigraphic structure and the geostress balance, and caused the stress to be rebalanced under the combined action of the geologic field, the stress field and the seepage field. Once the water inrush channel was formed and the water-Strong, it will lead to water inrush accident, which is the real reason for coal seam floor water inrush.

The damage degree of the excavation works to the strata and the damage caused by the hydraulic water are affected by many factors. There are coal seam depth, coal seam inclination, plowing length of mining face, mining technology, floor rock layer and so on.

3.1.5 *Deposited and Discontinuous Plaeo-Weathered Crust*

For the north China coalfields, with the increase in coal seam depth, the Austrian gray water hazard problem more and more prominent. As the Austrian gray water pressure, water-rich, easy to lead to catastrophic water burst accident.

The distribution of Ordovician limestone in the north China coalfields is wide, and the thickness of the sediments is generally thick in the middle and north. However, the Ordovician strata and the top of the Carboniferous strata covered with the existence of intermittent deposition, mainly due to the formation process, due to the sea into the sea retreat repeated, resulting in strata uplift after the Ordovician exposed surface, weathering makes the Ordovician strata. And, the thickness of the Ordovician strata is fissured, that is, the ancient weathering crust, the thickness of the rock is generally 10–30 m. When the crust subsided and then further accepted the deposition, the ancient weathering crust by the formation of the Ordovician clay, clay material, calcite and Benxi group of aluminum mudstone and other filling, resulting in poor permeability, with a certain water blocking performance [6].

Through the analysis of many exploration drilling and water inrush, the thickness of the ancient weathering crust at the top of the Ordovician coal field in different ore districts is calculated (Table 3.4).

Based on the above analysis, the main factors of water inrush from coal seam are established from five aspects: confined aquifer, seepage rock formation, geological structure, excavation and intermittent paleo-castylene crust. Because of the combined effect of many factors, this is also the basis for scientific prediction of coal seam floor water inrush (Fig. 3.3).

Table 3.4 Top of Ordovician paleo-weathered crust thickness of coalfields in north China

Ore district	Jiaozuo	Fengfeng	Hanxing	Feicheng	Huozhou	Weibei	Chenghe
Thickness of O_2 top aquiclude (m)	20–30	20	0–30	0–50	10–15	10–20	0
Filling characteristics	Fill fissure with clay	Fill fissure with clay or calcareous	Local filling	Clay fill the aquifer	Fill with late sediment	Filling	

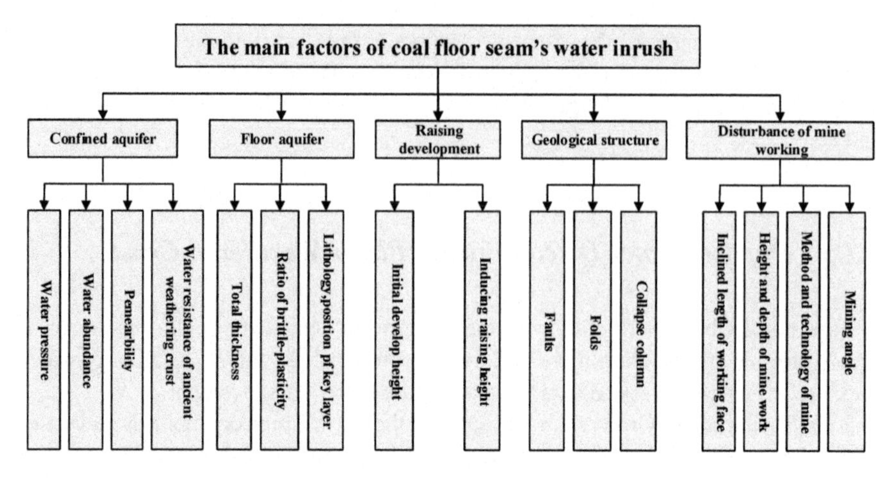

Fig. 3.3 The main factors of coal seam floor water inrush

3.2 Qualification of Main Control Factors

Through the analysis and collecting information of the index system of each main control factor, the indexes of each main control factor are quantified. Quantitative indicators should be accurately quantified when quantifying and qualitative indicators and semi-quantitative indicators should be quantified according to a certain standard.

3.2.1 Water Inrush Quantification

Main Indicators: quantification of water pressure, water-permeability, permeability and quantification of ancient weathered crust water barrier performance.

3.2.1.1 Quantification of Aquifer Water Pressure

Water pressure mainly depends on the pressure aquifer free water interface elevation and confined aquifer top interface elevation. The elevation of the interface of free water in confined aquifer can be collected through long-time dynamic observation hole of pressurized water or pumping (releasing) water and the elevation of the confined aquifer top interface can be acquired through borehole histograms. Water pressure is a quantitative indicator, and its quantification can be calculated from the water level measured by the hydrological hole (Eq. 3.1).

$$P = \rho g h \tag{3.1}$$

where P is water head pressure at the bottom interface; ρ is Groundwater density; g is gravity acceleration; and h is water barrier bottom interface to withstand the head height.

3.2.1.2 Quantification of Aquifer

There are many factors that affect the water-bearing capacity of confined aquifers, mainly including unit inflow, total aquifer thickness, mud consumption, core collection rate and so on. The water inflow from the collection unit can be tested by pumping (releasing) water, and the total aquifer thickness, mud consumption and core collection rate can be collected through the drilling histogram [7].

According to the provisions of the "Coal Mine Water Control" aquifer rich water, pumping (water) can be obtained by on-site test drilling unit water inflow value is divided into four levels.

- Weak water-rich: $q \leq 0.1$ L/s/m;
- Medium water-rich: 0.1 L/s/m $< q \leq 1.0$ L/s/m;
- Strong water-rich: 1.0 L/s/m $< q \leq 5.0$ L/s/m;
- Very rich in water: $q > 5$ L/s/m.

The unit water inflow in drilling units should be the standard unit water inflow when the aperture is 91 mm and the pumping depth is 10 m.

In addition, the hydrogeological exploration of most of the mining areas in our country is relatively low, and the experimental data of pumping (releasing) water are few. It is difficult to accurately determine the water-bearing capacity of the aquifer based only on the inflow of water in the drilling unit, so it is difficult to determine the water abundance of aquifer precisely only according to specific capacity of drill. To improve the evaluation accuracy, based on the analysis of several main controlling factors: aquifer thickness, specific capacity, core recovery rate, consumption of rinsing liquid, permeability coefficient, groundwater seepage fields formed by pumping tests and water inrush events (or gushing water), chemical field and geophysical analysis data affecting the water rich property of aquifer. Using multi-source information fusion theory, above factors controlling the water-bearing capacity of aquifers are quantified, dimensionless, weighted as well as we can use GIS to stack the control factors [8]. Determine the property of aquifer quantitatively using water-based index (Formula 3.2). And, using these four levels of drilling unit water inflow to test and correct water-rich results.

$$V_I = \sum_{i=1}^{n} V_i \cdot f_i(x, y) \tag{3.2}$$

where, V_I is watery index; V_i is the weight of the i factor affecting the watery; f_i (x, y) is the influencing value function of the i factors affecting watery; (x, y) are coordinates; n is the number of factors affecting watery.

3.2.1.3 Quantification of Aquifer Permeability

The permeability of aquifers is mainly measured by the permeability coefficient of aquifers. The main factors include the viscosity of aquifer medium and groundwater. The permeability coefficient K is a quantitative index. The permeability coefficient can be determined directly according to hydrogeological test (pumping (discharge) water test and water injection test) and so on in the hydrogeological survey of mining area.

3.2.1.4 Quantification of Plaeo-Weathered Crust Water Resistance

The paleo-weathered crust at the top of the Ordovician limestone has different water blocking properties according to the filling degree. In this paper, the effective water blocking thickness of the ancient weathering crust is used to quantitatively represent the water blocking performance of the ancient weathering crust, and the semi qualitative and semi quantitative method is used to determine the effective water blocking thickness of weathering crust. According to the different filling types of ancient weathering crust, the thickness of ancient weathering crust is converted into conversion thickness depends on the filling type, and the effective thickness is obtained by adding the thickness conversion coefficient of ancient weathered crust (Eq. 3.3). The conversion coefficient of the general filling type is 1, the conversion coefficient of the semi filling is 0.5, and the conversion coefficient of no filling is 0.25.

$$m = \sum_{i=1}^{n} \mu_i \cdot m_i \tag{3.3}$$

where, m is the effective water blocking thickness of the ancient weathering crust; μ_i is the corresponding conversion coefficient of the i-th charging section of the ancient weathering crust; and m_i is the corresponding thickness of the i-th charging section of the ancient weathering crust.

To improve the accuracy of evaluation, based on the analysis of several main control factors that affect the water resistance of the ancient weathering crust, such as the thickness of ancient crust, filling degree, core collection rate and flushing fluid consumption, and the water resisting performance of weathered shell is comprehensively determined by using the GIS based water resistance index method (Formula 3.4).

$$T_I = \sum_{i=1}^{n} t_i \cdot f_i(x, y) \tag{3.4}$$

where T_I is the water resistance index of the ancient weathering crust; t_i is the factor weight of i-th affecting the water resistance of ancient weathering shell; $f_i(x, y)$ is the i-th value of influencing factors of water resistance of ancient weathering shell;

(x, y) are the coordinates; n is the number of factors affecting the water resistance of ancient weathering shell.

3.2.2 *Quantification of Coal Seam Floor*

3.2.2.1 Quantification of Mining-Induced Fracture Zone in Coal Seam Floor

According to the theory of "down three zones" of coal seam mining, the coal seam floor will produce a certain depth of mine pressure damaging zone due to the combined effect of geological field, stress field and seepage field after coal mining. It has water conductivity. Factors affecting the depth of mine pressure damaging zone are: coal seam depth, long face, coal seam dip, coal seam thickness, mining technology and so on. The pressure damaging developing zone is mainly determined by many factors, collected through the excavation engineering layout map, drilling histogram, stratigraphic section and geological reports, and technologies such as field measurement, empirical formula, physical simulation or numerical simulation [9, 10]. Figure 3.4 shows the model of the "down three zones." The thickness of mining-induced fracture zone can be empirically calculated by:

$$h_1 = 1.86 + 0.11L \tag{3.5}$$

where h_1 is the thickness of the mining-induced fracture zone and L is the length of the working stope.

3.2.2.2 Quantification of Intact Rock Zone

This paper quantitatively determines the water-barrier performance of floor aquifer rocks based on the converted thickness of intact rock zone and the thickness and location of key stratum.

Quantification of the thickness of the complete rock zone: According to the theory of "under the three band", the method of quantitative determination of the converted thickness of the complete strata is divided into two steps:

Step 1: Using the total aquiclude thickness minus the top of the aquiclude at the top of the stratigraphic subsidence zone and aquiclude at the bottom of the strait development zone thickness, we can get the complete thickness of the lithosphere in the "next three belts" (Eq. 3.6).

$$h_2 = H - h_1 - h_3 \tag{3.6}$$

Step 2: Calculate the thickness of intact rock strata according to the comprehensive conversion coefficient in Table 3.3 according to different lithology strata, and add them to the converted thickness of complete strata (Eq. 3.7).

$$h_2' = \sum_{i=0}^{n} \mu_i h_{i2} \tag{3.7}$$

where h_2 is the thickness of intact rock zone; H is the total thickness of aquiclude; h_1 is the thickness of the damaged zone of mining floor; h_3 is the confined water to guide the thickness of the upper zone; h_2' is conversion thickness for intact rock zone; μ_i is lithology conversion factor; h_{i2} is thickness of i-th layer of brittle rock in complete lithosphere.

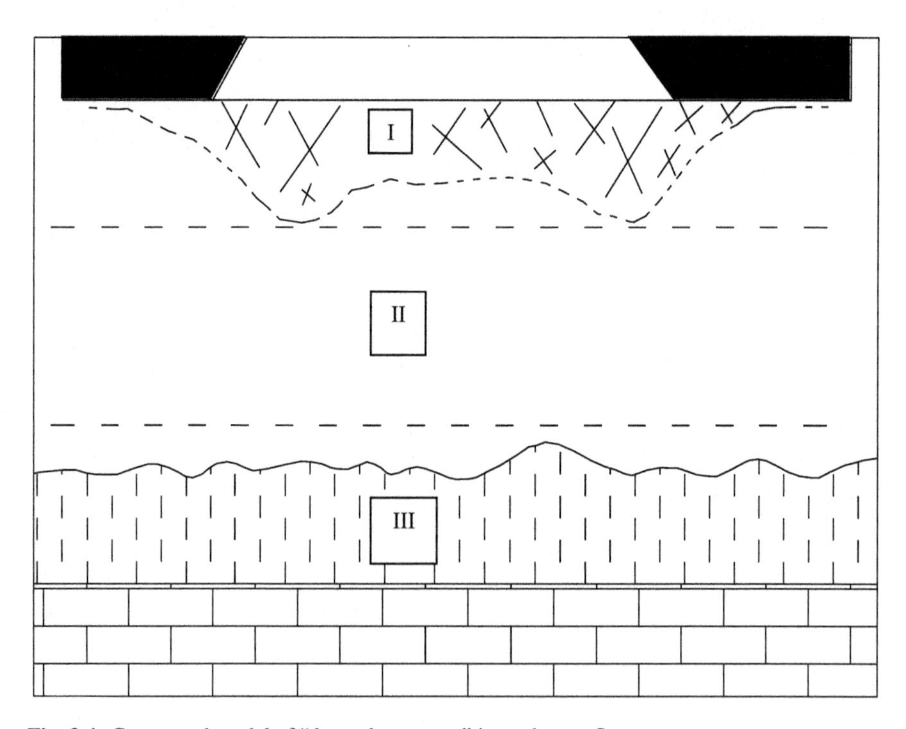

Fig. 3.4 Conceptual model of "down three zones" in coal seam floor

The thickness quantification of the key stratum in the water resistance of aquiclude is also related to its lithological assemblage, in addition to the aquiclude thickness and lithology. Only brittle and plastic rock are combined with each other, the water barrier was stronger. And only when the key stratum is within the complete lithosphere can its water-repellent performance be better [11–13]. The key stratum thickness in this paper is quantified by the method of accumulating the brittle rock thickness in the complete lithosphere, the key stratum thickness statistics formula is (Eq. 3.8), and the sum of the stratified statistics of the study section is the total thickness of the key stratum of the section.

$$M = \sum_{i=1}^{n} m_i \tag{3.8}$$

where, M is the thickness of the key rock formation; m_i is the i-th brittle rock thickness in the complete rock zone.

3.2.2.3 Quantification of Confined Water Conductivity

Confined water guide belt, also known as the lead development zone, that is the groundwater in the confined aquifer rises along the height of the bottom joint of the aqueduct or the fractured zone According to the engineering excavation activities on the joint or fracture with no secondary damage, is divided into the original lead before mining and mining induced high. Due to the different lithology and structure at the bottom of aquiclude, the original lead height of different mining areas varies greatly. And due to different mining technologies, face layout, coal seam thickness dip and pressure on working face, the mining induced Lead high is not the same, but its height is generally not large. Due to the above reasons, the upper boundary of the uplift zone is generally uneven. Even some mines may not develop zone due to the presence of paleo-weathering crust on top of the Ordovician limestone.

At present, the original lead height before mining and the induced lead height are mainly determined by field drilling, similar material simulation and numerical simulation. Confined water guide high-band quantization formula (Eq. 3.9).

$$h_3 = h_3' + h_3'' \tag{3.9}$$

where, h_3 is confined water guide high-band thickness; h_3' is the pre-mining lead high thickness; h_3'' is induced high-induced mining thickness.

3.2.3 Geological Structure

3.2.3.1 Quantification of Fault Distribution

Quantification of the fault distribution mainly considers the fault fracture zone and the width of fault influence zone (also called fault buffer zone or fault fracture zone). Based on experience, the width of the normal fault zone can be calculated by Eq. 3.10. According to the lithology of the two plates of the normal fault, the values of the coefficients related to the lithology of the normal faults are different [14–17]. According to experience, the correlation coefficient is 1.14 when the disks are soft rock and coal or other loose rock layers respectively. The correlation coefficient is 0.76 when the normal faults are all medium hard rock layers. When the normal faults are two hard rock layers, the coefficient is 0.38 [11]. The width of the normal fault fracture zone corresponds to half of the fault influence zone, as expressed in Eq. 3.11.

$$K_y = \gamma h^{3/5} \tag{3.10}$$

$$K_b = K_y/2 \tag{3.11}$$

where, K_y is the normal fault influence zone width; γ is the coefficient related to the lithology of both plates in the normal fault; h is the normal fault drop; K_b is the normal fault influence zone width.

When multiple layers are cut by normal faults, the weighted average of the fault influence width is taken (Eq. 3.12).

$$K_y = \frac{\sum_{i=1}^{n} \gamma_i m_i h^{3/5}}{\sum_{i=1}^{n} m_i} \tag{3.12}$$

where, K_y is the normal fault influence zone width; γ_i is the coefficient related to the lithology of both plates in the i layer of normal fault cutting; n is the number of rock layers cut by normal faults; m_i is the i-th layer thickness of the normal-fault cut; h is the normal fault drop.

As for the reverse fault fracture zone and the width of inverse fault influence zone, in this paper, according to the width of reverse fault influence zone is often lager than the normal fault. According to the statistical summary of multiple mining areas, the quantification of reverse fault influence zone can be calculated by referring to formula 3.10 or formula 3.11 and then multiplied by the coefficient of 1.2, that is Eq. 3.13. The ratio of the width of the reverse fault fracture zone to its influence zone width is 1: 2.5, by Eq. 3.14.

$$K'_y = 1.2\gamma h^{3/5} \tag{3.13}$$

$$K'_b = 0.4K'_y \tag{3.14}$$

where, K'_y is the reverse fault influence zone width; γ is the coefficient related to the lithology of both plates in the normal fault; h is the reverse fault drop; K'_b is the reverse fault influence zone width.

The risk of water inrush from fault distribution is mainly divided into two steps. The first step: Obtained the fault fracture zone and the width of the influence zone, then determine the scope of each fault fracture zone and influence zone according to the length of the fault in the study area. The second step is based on the different stress forms of normal and reverse faults, as well as the differences of fracture development between the normal and reverse fault zones and fracture zones. The fault distribution water inrush risk characteristics are respectively assigned to the influence zone and the fracture zone. According to the nature of fault and the location of fault, the value of the fault distribution water inrush risk characteristic is different, as shown in Table 3.5.

3.2.3.2 Quantification of Folds

The most easily formed part of the pleated water channel is the crimp section of the joints. There are two steps to the quantization of the fold distribution. The first step is to determine the width of the fold axis. In the same hydrogeological unit, the width of the anticlinal fold axis is the maximum of the width of the affected belt [18]. The width of the oblique fold axis is 60% of the corresponding anticline width, and the area of each fold axis is determined according to the length of the folding curve in the study area. The second step is to the study area fold crankshaft to give the water breakthrough risk characteristic value. Because the anticlinal axis is mainly affected by tensile stress, the axial section of the anticlinal axis is mainly subjected to compressive stress, and the joint fracture of the anticlinal axis is the development of the oblique axis, so anticline and syncline distribution characteristics of water inrush risk value assignment is different, as shown in the Table 3.6.

3.2.3.3 Quantification of Collapse Column

The fracture development area around the subsided column is called the influence area. The subsided column area and the influence area may become mine sudden

Table 3.5 Characteristic values of water bursting risk of fault distribution

Location in fault	Fracture zone of fault	Affected zone of fault
Assigned value	0.7	1

Table 3.6 Characteristic value of water bursting risk of fold distribution

Fold type	Anticline	Syncline
Assigned value	1	0.7

Table 3.7 Characteristic values of water bursting risk of collapse columns

Part of collapse column	Collapse body	Affected zone of collapse
Assigned value	1	0.8

(gushing) water channel. The distribution of subsided column is quantified in two steps. The first step is to determine the subsided column area and the influence area, the column area mainly depend on the geophysical or underground exposed measurements were obtained, while the area of influence area is the area of the annular zone of the length of one sixth of the major axis of the subsided column [19]. The second step is to distinguish the column area and the influence area from the subsided column in the study area to give water inrush risk characteristics. Because of the karst development in the subsided column area, the influence area is mainly the joint fracture development, so the assignment of the water inrush risk quantification coefficient is different in the column area and the influence area, as shown in Table 3.7.

The spatial development of the subsided column, the degree of rock fragmentation and filling condition in the column affect the water conductivity of the collapse column comprehensively. The development range of Subsided column, the degree of rock column crust and filling are comprehensively determined by means of the geophysical exploration results map, drilling results maps and stratigraphic profiles.

3.2.3.4 Quantification of Pressure Damaging Zone

The intersection of two fault endpoints and the fault intersection, is stress-concentrated area Generally, the fractures of rock mass are more developed than other parts of the fault, and are easier to become a mine water inrush (or gushing water) channel, therefore, the fault intersection, and the end point is also an important factor controlling coal floor water inrush [20]. Quantification of fault endpoint is divided into two steps. The first step is to determine the area of the water inrush (or gushing water) buffer zone at the fault, due to the stress concentration at the end point, the development zone of influence zone is larger than other parts of the fault, we call the area extending the width of the fault affected zone around the endpoint as the endpoint of water inrush (or gushing water) buffer. Its area can be quantified by Eq. 3.15. The second step is to assign water inrush risk characteristics to the fault endpoint of water inrush (or gushing water) buffer, valuing 1.7 uniformly.

$$S_2 = 2K_y(4K_y + K_b) \tag{3.15}$$

where, S_2 is the area of the fault endpoint of water inrush (or gushing water) buffer, K_y is the width of fault fracture zone, K_b is the width of fault affected zone.

The intersections of fault intersections are also quantified in two steps. The first step is to determine the area of intersection of two faults, each fault contains a crushing zone and two impacted zones, two faults intersect to form three types of regions, that is, one intersecting area of fracture zone and fracture zone, four intersecting areas of fracture zone and affected zone, four intersecting areas of affected zone and affected zone. According to the calculation method introduced in Sect. 3.3.5.1, the width of the fault fracture zone and affected zone can be calculated, and according to the actual situation of fault intersection, the area of three types of areas formed by the intersection of two faults can be determined. In the second step, the three types of regions formed by the intersection of faults are respectively given the water inrush characteristics, according to the intersection of the region is not the same as its water inrush characteristics of the value shown in Table 3.8.

3.2.3.5 Quantification of Distribution of Intersections Between Faults and Folds

When the fault intersects the fold, the danger of water inrush at intersections increases, therefore, the intersection of faults and folds is also an important control factor that affects the water inrush of seam floor. The intersection of fault and fold can be quantified in two steps. The first step is to determine the area of intersection of fault and fold axis, fault and fold axis intersect to form two types of regions, that is, one intersecting area of fault fracture zone and fold axis, two intersecting areas of fault affected zone and fold axis. We can calculate the width of the fault fracture zone and fault affected zone according to the method introduced in Sect. 3.3.5.1, and determine the width of the fold shaft as described in Sect. 3.3.5.2. According to the actual situation of the intersection of the fault and the fold, the area of the three types of areas formed by the intersection is determined. In the second step, the three types of regions formed by the intersection of faults and folds are respectively given water inrush characteristics, according to the intersection area is different, the water inrush risk characteristic values are shown in Table 3.9.

Table 3.8 Characteristic value of water bursting risk of fault intersections distribution

Fault type	Normal fault		Reverse fault	
	Fracture zone	Affected zone	Fracture zone	Affected zone
Assigned value	1.7	1.4	2	1.7

Table 3.9 Characteristic value of water bursting risk of fault and fold intersections distribution

Assignment	Fault			
	Fracture zone of normal fault	Affected zone of normal fault	Fracture zone of reverse fault	Affected zone of reverse fault
Anticline	2	1.7	1.7	2
Syncline	1.7	1.4	1.4	1.7

3.2.3.6 Quantification of Distribution of Intersections Between Faults and Subsided Columns

Both faults and subsided columns may become mine water inrush (or gushing water) channels, when the fault cuts the column, the danger of water inrush in subsided column increases. Therefore, in this paper, the intersection of fault and subsided column is also used as a water inrush factor to control coal floor. The intersection of fault and subsided column can be quantified in two steps [21]. The first step is to determine the area of intersection of the fault and subsided column, because the area of subsided column in the impermeable layer of the plane is much smaller than the distribution area of fault. The area of subsided column (sum of the area of column zone and the area of affected zone) is taken as the intersecting area of fault and subsided column when they intersect. The area of subsided column is determined according to the method introduced in Sect. 3.3.5.3. The second step is to give the characteristic value of water inrush risk of the intersection of fault and subsided column, and we value 2 uniformly.

3.2.3.7 Quantification of Fault Scale Index

The number, density and size of faults in the study area are indicators of regional fault development. In this paper, the quantification of the fault scale index is used to quantification. The sum of the product of the divides of all faults in a unit area and its length in the study area is the fault scale index (Eq. 3.16). According to the density of faults in the study area, the unit area can be divided into the same cell grid of 100 m × 100 m, 500 m × 500 m and 1000 m × 1000 m.

$$V = \frac{\sum_{i=1}^{n} l_i z_i}{S_6} \tag{3.16}$$

where, v is the fault scale index; l_i is the length of the i-th fault in the study area; z_i is the divide of the i-th fault; n is the number of faults in the study area; S_6 is the area of the cell grid.

3.3 Dimensionless Processing of Main Control Factor Index

The main control factors of coal floor water inrush in the above quantification can be classified into three kinds of indexes, which are quantitative index such as the permeability of aquifer, qualitative index such as the distribution of the intersection of faults and folds; and Semi-quantitative and semi-qualitative index such as the equivalent thickness of paleo-weathering crust shown in Table 3.10. Because the dimension is not consistent between the control factors, for example with is the quantitative index of water pressure of water-filled aquifer and conductivity high thickness of confined water, which some main control factors have no dimension, such as all the main control factors in qualitative indicators. Based on the above reasons, when we use of many factors on the prediction of water inrush of the coal floor, still need to quantify the main control factors of data after unified dimensionless processing, and elimination of the influence of single factor, to the main controlling factors can be composite superposition [22].

There are a lot of factors that affect the bottom plate of the coal floor, and there are some factors that are more important, and the more dangerous it is, the more dangerous it is to break the ground, and the more it is, the greater the depth of the subsurface, the more bad it is, the more bad it is, and the more likely it is to have a sudden accident, and we call it as the incentive factor. On the contrary, some factors such as the greater thickness of key stratum, the smaller the risk of water inrush, we call this kind of index factor as punishment. We mainly use the maximum

Table 3.10 Index classification of main control factors

Index type	Quantitative index	Qualitative indicators	Semi-quantitative and semi-qualitative index
Main control factors	Water pressure, water content, permeability, mining floor—belt, the guide belt of confined water	The distribution of faults, the distribution of folds, the distribution of subsided column, the distribution of fault intersection and endpoint, the distribution of fault and fold intersection, the distribution of fault and subsided column	The thickness of the intact rock band, the thickness key stratum, the equivalent thickness of paleo-weathering crust, fault scale index

Table 3.11 Dimensionless classification of main control factors

Dimensionless method	Maximum value method	Eigenvalue method	Minimum method
Main control factors	Water pressure, water content, permeability, mining floor—belt, the guide belt of confined water, the distribution of fault intersection and endpoint, the distribution of fault and fold intersection, the distribution of fault and subsided column	The distribution of faults, the distribution of folds, the distribution of subsided column	The thickness of the intact rock band, the thickness key stratum, the equivalent thickness of paleo-weathering crust, fault scale index

value method (3.17) for the dimensionless of the index value of the motivator. We adopt minimum value method (3.18) for the dimensionless of the index value of the penalty factor. For the dimensionless in motivating factors, for example, when the distribution of faults and folds and subsided column is used to quantify of assignment method, that its parameter values were no dimension, such processing method called characteristic value assignment method. The dimensionless classification of main control factor is shown in Table 3.11.

$$Y_i = \frac{y_i - y_{\min}}{y_{\max} - y_{\min}} \tag{3.17}$$

$$Y_i = \frac{y_{\max} - y_i}{y_{\max} - y_{\min}} \tag{3.18}$$

where, Y_i is the index value after dimensionless at I point; y_i is the quantitative index value of the influence factor at I point; y_{max} is the maximum quantitative index value in the research area for influencing factors; y_{min} is the minimum quantitative index value in the research area.

3.4 Summary

In this chapter, the main factors influencing the water inrush from coal seam floor are systematically analyzed. Methods to acquire, quantify and dimensionless process these main control factors are detailly introduced, and then main factors index system for water inrush from coal seam floor is established, laying foundation for multi-factors information fusion analysis.

References

1. Wu, Qiang, and Siyuan Ye. 2008. The prediction of size-limited structures in a coal mine using artificial neural networks. *International Journal of Rock Mechanics and Ming Sciences* 45 (6): 999–1006.
2. Lu, Juan, and Gang Chen. 2013. Research progress and Discussion on floor water inrush prevention and control in coal mine. *Western Resources* (2) (in Chinese).
3. Bailey, W.R., J.J. Walsh, and T. Manzocchi. 2005. Fault populations, strain distribution and basement fault reactivation in the EastPennines Coalfield, UK. *Journal of Structural Geology* 27: 913–928.
4. Santos, C.F., and Z.T. Bieniawski. 1989. Floor design in underground coalmines. *Rock Mechanics and Rock Engineering* 22 (4): 249–271.
5. Jorge, M., S. Javier, and J. Ruben. 2002. Numerical modeling of the transient hydrogeological response produced by tunnelconstruction in fractured bedrocks. [J]. *Engineering Geology* 64 (4): 369–386.
6. Arguello J.G, C.M Stone, and J.C. Lorenz. 1996. Geomechanical numerical simulations of complex geologic structures. In: *Rock Mechanics*, ed. M. Aubertin, and F. Hassani, vol. 2, 1841–1848. Rotterdam, AA. Balkema.
7. Wang, Jingming, Jiade Ge, Yuhua Wu, et al. 1996. Mechanism on progressive intrusion of pressure water under coal seams into protective aquiclude and its application in prediction of water inrush. *Journal of Coal Science & Engineering* 2 (2): 9–15.
8. Wang Jingming. 2004. *Water inrush mechanism and its application of progressive upward lifting of confined water along coal seam floor*. General Research Institute of coal science (in Chinese).
9. Zhang Jincai, Tianquan, Liu. 1990. On the depth and distribution characteristics of mining fracture zone in coal seam floor. *Journal of China Coal Society* 15 (2): 46–55 (in Chinese).
10. Xu, Xuehan, and Jie Wang. 1991. *Prediction of water inrush in coal mine*. Beijing: Geological Publishing House (in Chinese).
11. Li, Kangkang, and Chengxu, Wang. 1997. Rock mass in-situ testing technology for study of water inrush mechanism of coal seam floor. *Coal Geology and Exploration* (3): 31–34 (in Chinese).
12. Saaty, T.L. 1977. A scaling method for priorities in hierarchical structures. *Journal of Mathematical Psychology* 15: 234–281; Saaty, T.L. 1980. *The analytic hierarchy process*. McGraw-Hill, New York, 112–155.
13. Wu, Q, J. Wang, D. Liu, F. Cui, and S. Liu. 2007. The new practical evaluation method of floor water inrush. IV: Application of AHP type vulnerability index method based on GIS. *J Chin Coal Soc* 34 (2): 233–238.
14. Wu Q, S. Xie, Z. Pei, and J. Ma. 2007. The new practical evaluation method of floor water inrush. III: Application of ANN type vulnerability index method based on GIS. *J Chin Coal Soc* 32 (12): 1301–1306.
15. Wu, Q, Z. Zhang, S. Zhang, and J. Ma. 2007. The new practical evaluation method of floor water inrush. II: The vulnerability index method. *J Chin Coal Soc* 32 (11): 1121–1126.
16. Feng, Dawei. 2010. *Study on prediction and prevention of coal mine water disasters in Chongqing Nanchuan*. Guiyang: Guizhou University (in Chinese).
17. Xiao, Youcai. 2013. *Mechanism and engineering practice of outburst and seepage induced by water inrush from coal seam floor*. Xuzhou: China University of Mining and Technology press (in Chinese).
18. Guan, Entai. 2012. *Origin of water bursting coefficient and process of modification. Coal Geology of China* 24 (2): 30–32 (in Chinese).
19. Longqing, Shi, and Han Jin. 2005. Theory and practice of dividing coal mining area floor into four-zone. *Journal of China University of Mining and Technology* 42 (1): 16–23. (in Chinese).
20. Niu, Jianli. 2008. *A study on couplingeffect beween rock and water in coal floor and safety mining technology under high groundwater pressure*. Doctoral dissertation of Xi'an Branch of Coal Science Research Institute (in Chinese).

21. BAI, Haibo. 2008. Seepage characteristics of top stratum of ordovician system and its application study as key aquifuge. Xuzhou: China University of Mining and Technology (in Chinese).
22. Liu, Weitao, Wenquan, Zhang, and Jia, Li. 2000. An evaluation of the safety of floor water-irruption using analytic hierarchy process and fuzzy synthesis methods. *Journal of China Coal Society* 25 (3): 278–282 (in Chinese).

Chapter 4
Vulnerability Index Method Based on Partition Variable Weight Theory

4.1 Information Fusion Model of Water Inrush Evaluation

4.1.1 Definition of Information Fusion

Multi-source information fusion is prevalent in nature. The application of information fusion technology in engineering field is the imitation and abstraction of biometric information fusion function by using mathematical methods and computer technology. In the 1970s, the term "data fusion" firstly appeared in the military field for the construction of military C3 (command, control, communications and intelligence) system. That is, the embryonic form of information fusion jointly processed all the data measured by multi-sensor of C3 system to obtain more accurate and effective judgments. By the end of the twentieth century, information or the source of data began to become diversity, including data, signals, and even knowledge, symbols and experience, and so on. At this time, many concepts of information fusion in line with the characteristics of this research field are given by different research directions and engineering applications field. After more than half a century of development, the current definition of information fusion is generally defined as: it is a series of information processing processes with the use of modern mathematical methods and computer technology to registrate, reorganize, associate and optimize, under a certain standard, multi-source information in many ways and multi-level to achieve mission tasks [1–4].

It can be seen from the definition of information fusion that multi-source and multi-style information is the foundation and the object to be processed, and multi-aspect and multi-level registration, reorganization, association and optimization are its core. Modern mathematics and computer technology are its useful methods and tools. Effective and useful decision-making are the goal of information fusion technology.

Compared with the traditional single data analysis and processing decision, the advantages are compared with: expanding the time domain and spatial domain to obtain the information data range, extracting and registering the associated information data; improving the effective utilization and status of the information data The accuracy of the evaluation, the level of the integration between the various factors within the level of consideration in the evaluation of the importance of the

Y. Zeng, *Research on Risk Evaluation Methods of Groundwater Bursting from Aquifers Underlying Coal Seams and Applications to Coalfields of North China*, Springer Theses, https://doi.org/10.1007/978-3-319-79029-9_4

target evaluation to reduce the degree of fuzzy reasoning; verification process of the integration system monitoring, enhance the integration of adaptive systems to optimize the system-wide performance. The development of computer technology and mathematics has also promoted the application of information fusion technology in more industries and fields, namely, traditional agriculture, national defense and transportation, including geoscience, meteorological forecast, economy and medical care. The research of information fusion mainly includes the research of the whole fusion architecture design, the hierarchical process design, the fusion algorithm at each level, and the fusion algorithm of the factors within the hierarchy.

4.1.2 Category and Models of Information Fusion

Data processing, hierarchical identification and reasoning decision are three major processes of information fusion technology. Different ways of information fusion, classification is also diverse. Such as according to the abstraction process to deal with the abstract level of the object, it is divided into data, features and decision-level integration Ref. [22] in Chap. 3. A summary of the three levels of fusion is shown in Fig. 4.1.

Data-level fusion is shown in Fig. 4.1a, which is a low-level fusion. The main way is to extract, register and correlate the observed information data. And based on the results of the fusion layer to extract the characteristics of decision-making. This

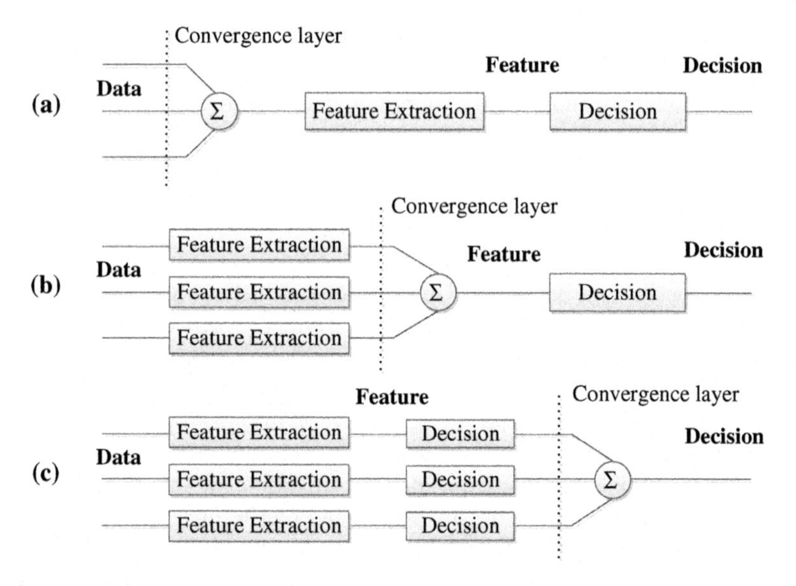

Fig. 4.1 Three levels of information fusion, **a** Data Level Fusion; **b** Feature Level Fusion; **c** Decision Level Fusion

fusion is oriented to the information data itself, the utilization of the fusion object is high, but because the relationship between the information cannot be considered, and the fusion processing is less time-dependent.

Feature level fusion is shown in Fig. 4.1b, which is a fusion of intermediate levels. The difference between the above and the above fusion is that the multi-source information data processing requires not only the extraction, registration and association of the information data, but also the feature extraction and attribute description of the information data. And then the system integration center of the extracted feature vector and attribute fusion processing. The lack of processing is that some useful data information will be lost, reducing the fusion performance, low accuracy. There are artificial neural network method, analytic hierarchy process, K nearest neighbor method and clustering method in hierarchical process design, hierarchical and hierarchical factors.

Decision-level integration is shown in Fig. 4.1c, which is a combination of high-level integration and appeals. First, a single sensor uses the respective information data to make the initial decision, the fusion system re-extraction, registration and association of the initial decision-making, by the system integration center to decision-making optimization and reasoning. Diversity of data on the level of integration of small interference, the system integration center less. The disadvantage is that the loss of information data, data utilization is small, low accuracy. Hierarchical process design, hierarchical and hierarchical factors within the integration of the main algorithms, expert system, D-S evidence reasoning.

In recent decades, in the field of information fusion and engineering applications, there have been many fusion models in line with their fields, various models of fusion architecture, and hierarchical process design. Such as JDL model, Dasarathy model and waterfall model. One of the most typical, now the most mature, popular or JDL model, it is the 20th century, 80 years by the US Defense Data Fusion joint command laboratory first proposed, the overall model of the model shown in Fig. 4.2. JDL model is divided into four stages to carry out information evaluation, that is, three assessments and an optimization cycle.

The source optimization is a simple classification of the information measured by the multi-sensor to each processing unit, which is prepared for the next step.

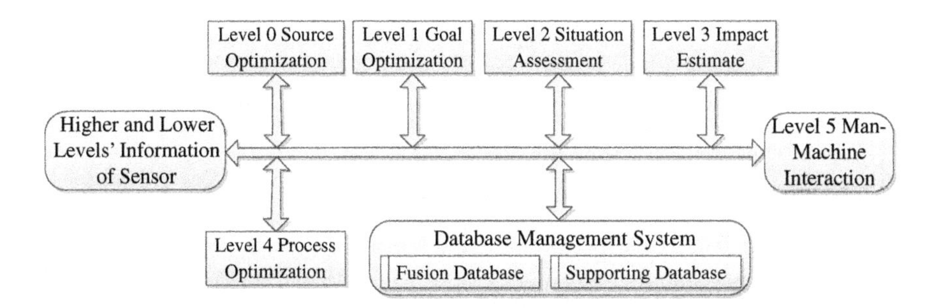

Fig. 4.2 The JDL model

The first stage is the target assessment. The location, parameters, identification information combination, access to the precise expression of each object, the formation of the object location, parameters, identification information and other attribute information database. Mainly to complete the four key functions: the sensor data into a unified unit and coordinate system, that is, data calibration; the sensor data assigned to each object, that is, data/object association; timely estimate the object location, motion characteristics and attributes; object recognition and classification.

The second stage is the assessment of the situation. The information objects obtained by the target evaluation are aggregated according to one way, the trend graph of the fusion framework is determined, and the behavior of the whole situation is assessed by the database information.

The third stage is a threat assessment. Analyze the outcome of the situation assessment based on the participant's vision or predicted behavior, map from the previous stage to the future, and analyze the pros and cons of future scenarios or behaviors.

The fourth stage is process assessment (or process optimization). Process evaluation is a cycle identification and verification optimization process for the whole system model, which evaluates system performance during the process of information fusion. Through the specific optimization indicators, identify new and valuable information, the entire system real-time monitoring, evaluation and tuning.

4.1.3 GIS-Based Information Fusion Model of Floor Water Inrush Evaluation

From the analysis of the previous chapter, it can be seen that many main factors influence the water inrush from coal seam floor, and the water inrush mechanism is complicated. Multiple influencing factors and their importance must be taken into account to evaluate water inrush. In view of the above analysis, we will be a variety of factors affecting the floor water inrush as JDL fusion model data input, and the impact of information on the use of GIS calibration storage, correlation processing and identification display process operation, and the use of modern mathematical methods of science To determine the relative importance of the influencing factors on the water diversion of the size of the index, that is, weight, and thus establish a GIS-based floor water inrush evaluation JDL-type information fusion model, as shown in Fig. 4.3.

[1] Information preprocessing. Also in the source optimization, in the coal exploration, construction and production process, the accumulation of many coal-bed floor water burst-related multi-source geoscience information. (Water level, water temperature, water chemistry) information, field test data, laboratory test data and various expert experience, etc.; the above information data generally have the same time sequence inconsistency, the amount of information Gang is not uniform, both qualitative text and quantitative measurement data. Because of the types and forms of information data, a large amount of basic information

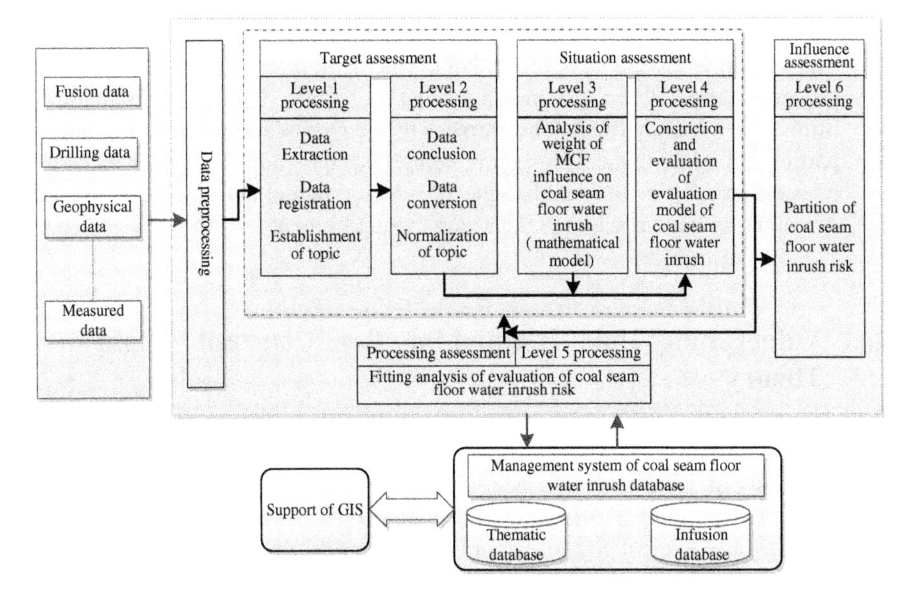

Fig. 4.3 Model of information fusion of coal floor water bursting

data needs to be analyzed, including the formation of coal seam floor water and water level of the water level line and other pretreatment.

[2] Target assessment. Model 1 and 2 of the model. According to the information preprocessing results, the preliminary estimation and reasoning of the evaluation objectives are the basis of the following situation assessment and risk assessment. The preliminary estimation and reasoning include: determining and extracting the data of each main control factor, registering the information in the space-time domain and the spatial domain, and establishing the thematic layer of the main control factor. Based on further analysis of the main control factors, dimensional unity and standardization of changes, the establishment of normalized thematic layer.

[3] Situation assessment. The level 3rd, 4th, processing in model 4.3. Based on the target assessment ratiocinate the complex relationship between the information data. The reasoning process of information relations includes: quantitatively determining the relative relationship between the influencing factors by introducing the concept of scale, and quantify the evaluation factors, determine the relative importance of the main control factors to the water inrush, and establish mathematical model of relative importance to floor water inrush.

[4] Hazard assessment. The level 6 processing in model 4.3 is a risk assessment. Based on the fitting analysis of the situation assessment results, the risk assessment of the water inrush from the bottom of the coal seam is made, and evaluate specific results according to partition, the targeted countermeasures and suggestions to prevent water are given.

[5] Process assessment. The level 5 processing in model 4.3 is a process assessment. The process assessment is a process of cyclic identification and an optimization process of the whole water inrush evaluation system. Using the vulnerability fitting rate to verify the risk assessment of the risk of water inrush evaluation partition, if it fails to meet the requirements, it needs to adjust and recalculate the relevant parameters, re-establish the JDL fusion model until the fitting results of evaluation model achieves the accuracy requirements.

4.2 Vulnerability Index Method Based on Constant Weight Theory

In the JDL-type information fusion model of floor water inrush evaluation based on GIS, the most critical level of treatment is the level 3 assessment under the situation assessment, that is, using a variety of mathematical methods to determine the weight between the various main control factors. Mathematical methods to determine the weight can generally be divided into two categories: linear and non-linear. Linear method mainly includes: Analytic Hierarchy Process (AHP), etc.; nonlinear mainly include: artificial neural network (ANN), Logistic regression and evidence of law and so on. In the JDL-type information fusion model of floor water inrush evaluation based on GIS, which is the evaluation model of coal seam floor water inrush vulnerability, according to the different mathematical method to determine the weight, the vulnerability index method can be divided into GIS-AHP type, ANN type, Logistic regression method and evidence right type. Now ANN vulnerability analysis method and the AHP type vulnerability index method based on GIS are briefly analyzed.

4.2.1 Design of Weight Model

Quantitative determination of relative importance inside influencing factors and between influencing factors of water inrush coal seam floor. It is also the focus of the research of many scholars. The traditional method of forecasting and predicting water damage is generally expressed by the simple linear function relationship between the internal factors of water inrush and their relative importance, which cannot truly reflect the complex mechanism and evolution process of the main control factors. The artificial neural network and the analytic hierarchy process are a kind of mature method to deal with the multi-influencing factors weight.

4.2.1.1 Artificial Neural Network

According to the structure of the neural network and mechanized neural network, it has its important characteristics in information processing. The information processing depends on the synergistic completion of the neurons. The information storage and learning functions are realized by the change of the connection intensity. And information processing and storage share a unit; adjust the weights according to the changes in the information conditions or learning. Make the network output meet the specified target; can handle fuzzy, simulated and random information. And because of the diversity of geological information, information limitations and the complexity of the geological effect, ANN technology has been widely used in the field of geoscience.

(1) **ANN model**

Because AAN has similar characteristics with biological brain's nervous system, it has advantages in many information decision-dealing areas. ANN model is not the same, with its different functional characteristics, the scope of application is not the same. The choice of model is also changed according to different research objectives. Table 4.1 lists some constant weights used models and their applications [5–8].

(2) **BP neural network structure**

Here is a BP neural network model, which is a reverse transmission of multi-layer mapping network, and can correct the error, is currently the most widely used neural

Table 4.1 Features of ANN model

ANN model	Application area	Description
ADALINE	Recognition of model	Single-layer forward feedback network, generally used for pattern recognition
BP	Prediction of time or space	This is a classic multi-layer feedback network
HOPFIELD	Associative memory	Image recognition and so on
BAM (Bidirectional Associative Memory)	Bidirectional memory	Bidirectional search
BOLTZMAN	Optimization (shortest path) problem	An improved version of HOPFIELD
CP	Visual field	Belong to the competition network
SOM (Self-Organizing Map)	Control system	Is a self-organizing competition network, which can form a topology prediction structure between the input layer and the output layer
ARTl (Adaptive Resonance Theory)	Brain model	Mainly used to demonstrate the basic characteristics of network feedback principle

network. The essence of the BP neural network model is the feedforward network. There is no interconnection between the neurons at the same level of the network that consist of this network. The interaction between the same hierarchical elements is controlled by all the units of the next level. Neurons are connected directly from one layer to another, and the basic structure of this neural network is shown in Fig. 4.4.

(3) **BP neural network algorithm**

As can be seen from Fig. 4.4, this neural network is the mathematical relationship between multipoint input and multipoint output, which overcomes the relationship between multiple input and single output in other models. The following describes the principle of the BP neural network model.

[1] First set the initial weight $W(0)$, it is a small random non-zero value.
[2] According to the given input and output sample pairs, calculate the output y_i of the network, and then calculate the objective function of the network, according to the objective function and the size of the given fixed value ε to arrange. When the signal x_i is input, the neurons in the hidden layer are summed after weighting ω_{ji}, and then calculate the sum function f.

$$y_i = f(\sum \omega_{ji} x_i) \tag{4.1}$$

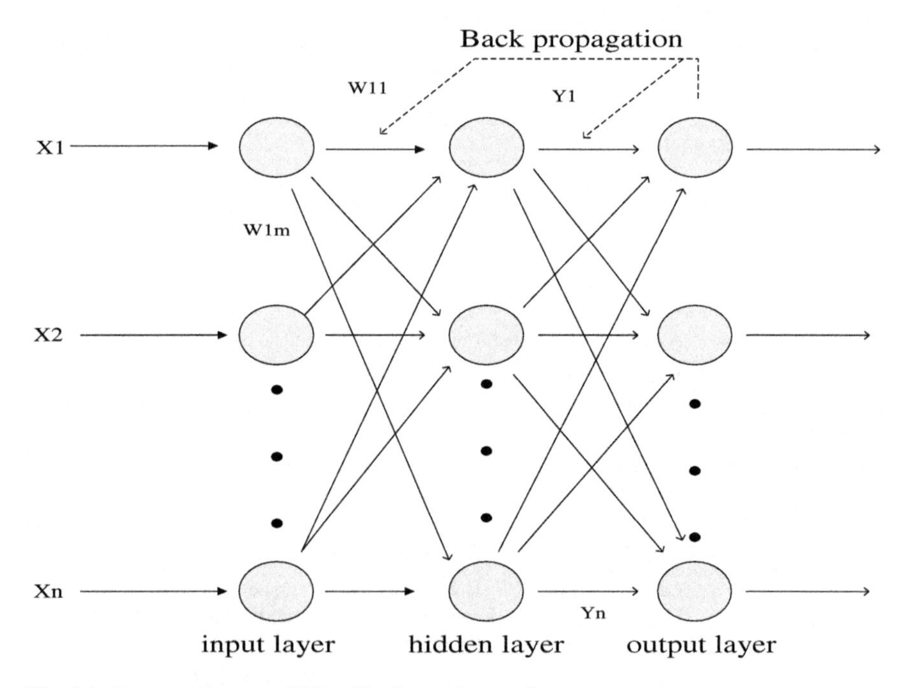

Fig. 4.4 Structure diagram of BP artificial neural network

Similarly, the output of the neurons in the output layer, the reverse propagation method gives the changes in weights $\Delta\omega_{ji}$ in neurons i and j, the expression is:

$$\Delta\omega_{ji} = \eta\delta_j x_i \tag{4.2}$$

where: η—learning rate parameters;
δ_j—depending on whether the neurons are the coefficients of the output neurons or implicit neurons

For output neurons:

$$\delta_j = (\frac{\partial f}{\partial net_j})\left[y_j^{(t)} - y_j\right] \tag{4.3}$$

For implied neurons:

$$\delta_j = (\frac{\partial f}{\partial net_j})\sum_q \omega_{qj}\delta_q \tag{4.4}$$

where: net_j—The weighted sum to neurons j of the input signals;
$y_j^{(t)}$—the target output of the neuron j reverse propagation calculation

[3] In the output layer, according to the objective function, according to the gradient descent method reverse calculation, it can be adjusted in layers.
In order to speed up the training, it often adds a momentum item onto the basis of the original, so that the previous changes in the right can affect the new weight changes, that is,

$$\Delta\omega_{ji}(k+1) = \eta\delta_j x_i + \mu\Delta\omega_{ji}(k) \tag{4.5}$$

where: $\Delta\omega_{ji}(k+1)$—the change of weight;
$\omega_{ji}(k)$—previous weight
$(k+1)$ and k—iteration numbers
μ—Momentum coefficient

4.2.1.2 Analytic Hierarchy Process

In 1974, the United States Professor Sade for the first time put forward the concept of hierarchical decision-analyzing method, and then it is simplified to Analytic Hierarchy Process (AHP). It is a kind of non-qualitative and non-quantitative fusion of scientific decision-making methods for the imperfection, uncertainty and non-linearity of natural information.

The central idea is to decompose the evaluation decision object into the main control factors, and divide the main control factors into groups and grading according to the principle of subordination, that is, construct a hierarchical hierarchy model. In this process, use human subjective analysis express non-quantitative factors in the form of quantitative. Between each level through the introduction of a reasonable scale, the use of pairwise comparison quantitative to determine the relative important weight ratio of the divisor in the same level. At the same time, it is necessary to construct the judgment matrix according to the level to calculate the weight ratio of each divisor of the next level in the upper level, and pass the influence onto the next level. In the hierarchical structure, the weights of each divisor in each level are determined, and the influence is passed to the next level, and the comprehensive weight of all divisor is achieved (total ranking). Finally, the research goal is decided using comprehensive weight. The application calculation process of AHP (Fig. 4.5) is as follows:

[1] Construct a hierarchical analysis model (Fig. 4.6). Analyze the research objectives, determine the main control factor and determine the membership function between the main control factors. According to the membership function, the target, the factor and the sub-factor are connected with the hierarchical struc-

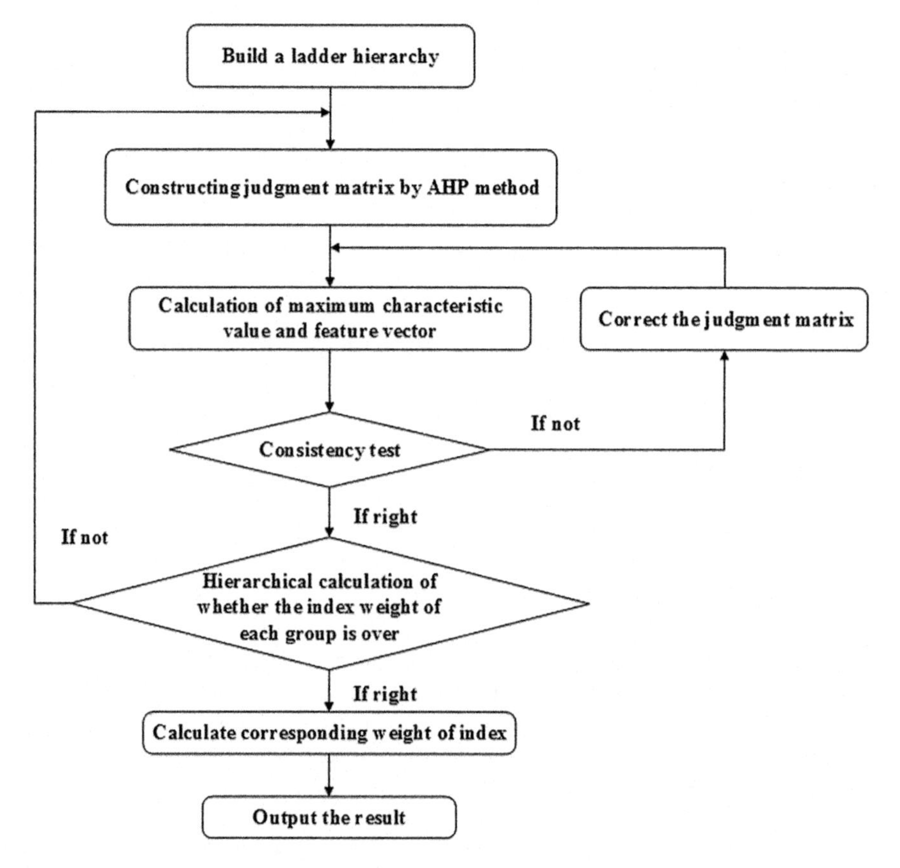

Fig. 4.5 Flowchart on the calculation of hierarchical structure

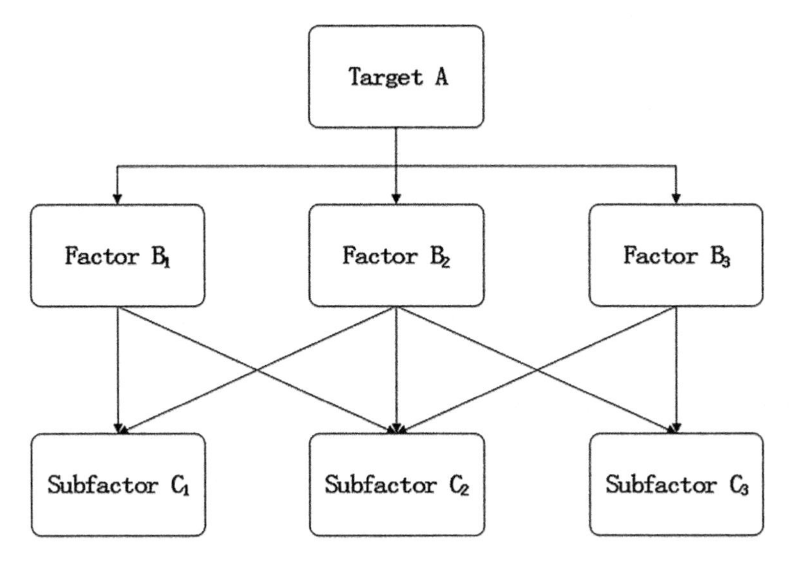

Fig. 4.6 Model of the AHP hierarchical structure

ture, and the hierarchical structure analysis model is constructed. The model generally includes the target layer (A level), the criterion layer (B level) and the decision layer (C level).

[2] Build the hierarchical judgment matrix. Between the hierarchy through the 1-9 scale concluded by Sade professor (Table 4.2), or the improved Sade 1-9 scale method (Table 4.3), to determine relative importance of the factors pairwise comparison quantity. The use of pairwise comparison of quantitative to determine the weight of relative importance of various level. The importance of the index $Y = \{y1, y2, \ldots, yn\}$ for the target T is that $(w_1, w_2, w_3, \ldots \ldots, w_n)$, $A = (a_{ij})_{n \times n}$ is the judgment matrix for all the factors pairwise comparison. This step is the key to use the scale method to quantify target evaluation.

$$A = \begin{bmatrix} w_1/w_1 & w_1/w_2 & w_1/w_3 & \ldots & w_1/w_n \\ w_2/w_1 & w_2/w_2 & w_2/w_3 & \ldots & w_2/w_n \\ w_3/w_1 & w_3/w_2 & w_3/w_3 & \ldots & w_3/w_n \\ \ldots & \ldots & \ldots & \ldots & \ldots \\ w_n/w_1 & w_n/w_2 & w_n/w_3 & & w_n/w_n \end{bmatrix}$$

[3] Hierarchical single factor ordering and consistency test. Calculate the matrix $A = (a_{ij})_{n \times n}$, that is, get the weight of each factor of the next hierarchy in the last hierarchy, where λmax is the largest eigenvalue of A and W is the corresponding eigenvector. Consistency ratio CR that summarized by professor Sade is used on the matrix A, and the corresponding value of the Sade 1–9 scale to RI is shown in Table 4.4; the improved 1–9 scale corresponding value to the RI value is shown

Table 4.2 Scale and context of Satty

Scale	Implication
1	Under Pairwise comparison, they have the same importance
3	Under Pairwise comparison, the former is slightly more important that the later
5	Under Pairwise comparison, the former is obviously more important that the later
7	Under Pairwise comparison, the former is strongly more important that the later
9	Under Pairwise comparison, the former is extremely more important that the later
2, 4, 6, 8	Represents the intermediate value of the adjacent judgment
Reciprocal	If the importance ratio of element i and element j is a_{ij}, then the ratio of importance of element j and element i is $a_{ij} = \frac{1}{a_{ij}}$

Table 4.3 Improved scale and context of Satty

Scale	Implication
5/5	Under Pairwise comparison, they have the same importance
6/4	Under Pairwise comparison, the former is slightly more important that the later
7/3	Under Pairwise comparison, the former is obviously more important that the later
8/4	Under Pairwise comparison, the former is strongly more important that the later
9/1	Under Pairwise comparison, the former is extremely more important that the later
5.5/4.5, 6.5/3.5, 7.5/2.5, 8.5/1.5	Represents the intermediate value of the adjacent judgment
Reciprocal	If the importance ratio of element i and element j is a_{ij}, then the ratio of importance of element j and element i is $a_{ij} = \frac{1}{a_{ij}}$

Table 4.4 The value of the average random consistency index RI

n	1	2	3	4	5	6	7	8	9
RI	0	0	0.58	0.90	1.12	1.24	1.32	1.41	1.45

in Table 4.5. When CR < 0.1, the weight meets the qualification; otherwise, the judgment matrix is reconstructed.

$$CR = \frac{CI}{RI} \tag{4.6}$$

$$CI = \frac{\lambda\max - n}{n - 1} \tag{4.7}$$

Table 4.5 The value of the average random consistency index RI corresponding to the improved scale

n	1	2	3	4	5	6	7	8	9
RI	0	0	0.169	0.2598	0.3287	0.3694	0.4007	0.4167	0.4370

where: *CI*—coincidence indicator of judging matrix
RI—random and average coincidence indicator.

[4] Total taxis of hierarchy and consistency test. Comprehensively consider the hierarchical levels, the weight of each factor to its indirect target layer is obtained (Table 4.6). And the consistency test is carried out with reference to step (3). Assuming that the Q layer is the upper target layer of the A layer, the A layer is the upper target layer of the B layer, and the single order weight of A to Q is a_1, a_2, …, a_m; the single order weight of B to A is :b_{1j}, b_{2j}, …, b_{nj} (bij = 0 when there is no association between Bi and Aj); then the weight of the B layer to the indirect target layer Q is $b_i = \sum_{j=1}^{m} b_{ji}a_j$ (i = 1, 2,…, n). The total ranking consistency ratio is shown in Eq. 4.8. When CR < 0.10, the overall sort consistency is acceptable, and the weight of each factor to the indirect target layer is determined.

$$CR = \frac{CI}{RI} = \frac{\sum_{j=1}^{m} CI_j a_j}{\sum_{j=1}^{m} RI_j a_j} \tag{4.8}$$

where *CR* is the consistency ratio of total taxis.

Table 4.6 The total sort of the hierarchy

B layer	A layer				
	A_1	A_2	…	A_m	Total taxis of B layer
	a_1	a_2	…	a_m	
B_1	b_{11}	b_{12}	…	b_{1m}	$\sum_{j=1}^{m} a_j b_{1j}$
B_2	b_{21}	b_{22}	…	b_{2m}	$\sum_{j=1}^{m} a_j b_{2j}$
…	…	…	…	…	…
B_n	b_{n1}	b_{n2}	…	b_{nm}	$\sum_{j=1}^{m} a_j b_{nj}$

4.2.2 Vulnerability Evaluation Method Based on Constant Weight Theory

4.2.2.1 ANN Type Vulnerability Index Method Based on GIS

Use GIS to conduct geological information processing and graphic expression. First, a variety of geological information is extracted, and then the information data is further quantized and normalized, and stored in the GIS attribute database. The information data processed by the GIS is used as the learning and training input of the ANN network. ANN expresses local factor in a unit, with the level of the factors are independent of each other. The interaction relationship is regulated and controlled based on the all units of next level, and the nonlinear relationship is used to quantified determine the mutual importance between the factors. The prediction results of artificial neural network (ANN) are used as input of GIS, use GIS's function of graphics overlay processing, and the prediction results of ANN are expressed in the way of graphic. The coupling of GIS and ANN not only can deal with the massive data information, but also uses non-linear thinking to deal with the complex relationship between multiple factors. And the use of GIS graphics output makes the decision more intuitive.

The basic idea of GIS and ANN coupling technology:

[1] Using GIS to preprocess the geological information, including information collection, quantization, data conversion and calculation. The establishment of GIS spatial database and the corresponding attribute database;
[2] Establish an ANN analysis model which can analyze the data comprehensively, and the model runs outside the GIS;
[3] The data needed for the model operation are taken from the spatial database and the attribute database of the GIS. The decision result of the coupled model is expressed by the mapping function of GIS, and the data information is stored in the GIS attribute database.

4.2.2.2 AHP-Based Vulnerability Index Method

By analysing the geological factors related to water inrush, the main control factor system which affects the permeation of the bottom plate is established. GIS is used to collect and quantify the main factors, and the single factor map and single factor attribute database are established by GIS. The relative importance of the main control factors on the floor water inrush is the key of fusion evaluation. The weight of each master is determined by AHP. And then the application of GIS-based information fusion technology for the normalization of the main control factors after the information fusion, and evaluation of the district [27–32]. GIS and AHP fusion evaluation of the basic ideas are as follows:

[1] Based on the analysis of hydrogeological conditions and the conditions of mine water filling, the main influencing factors affecting the permeability of coal seam floor are determined.

[2] After determining the main control factors, we need to extract the relevant data of various influencing factor from the collected data, and quantify the process, the quantitative process has two main aspects: First, conduct interpolation processing on collected data using Surfer software, to reflect the trend and distribution of a factor in the entire study area, that is, from point to surface to expand; Secondly, after using Surfer software to interpolate, the file after being interpolated need to be inserted into ArcGIS, the use of GIS spatial analysis and attribute calculation function are used to further processing, this process can be achieve quantitative spatial assignment, that is accurately described the value of a certain factor in the study area within any point.

[3] After quantifying the various factors, respectively establishing the thematic layers, it is needs to use AHP software to determine the impact of various factors on the floor water inrush. With pairwise comparison of various main control factors, to determine the degree between the factors in the form of marking, and thus establish a judgment matrix, in the premise that established judgment matrix is meet the consistency, we can finally calculate the respective weight of various impact in the floor water inrush.

[4] Because different factors have their own dimensions, in order to put different dimensions of the factors together to deal with the calculation, it is need to nondimensionalize each topic layer respectively, to eliminate the impact caused by the dimension. Then use the "superposition" function of GIS to "integrate" the thematic layers into a layer, and the attribute calculation is carried out according to the weight of each influencing factor, and finally the vulnerability index of each "segmentation unit" is obtained.

[5] Based on the size of the vulnerability index and the frequency of occurrence, the "Natural Breaks" is used to determine the partition threshold for each segment to achieve the final evaluation partition (Fig. 4.7).

4.3 Vulnerability Index Method Based on Partition Variable Weight Theory

4.3.1 An Overview of Variable Weight Theory

4.3.1.1 Significance of Variable Weight Theory in Floor Water Inrush Evaluation

Variable weight theory is a kind of technology to process index weight of factors, which is the important method of factor change theory. The core idea is based on different combinations of environment, when a factor index value changes, the weights

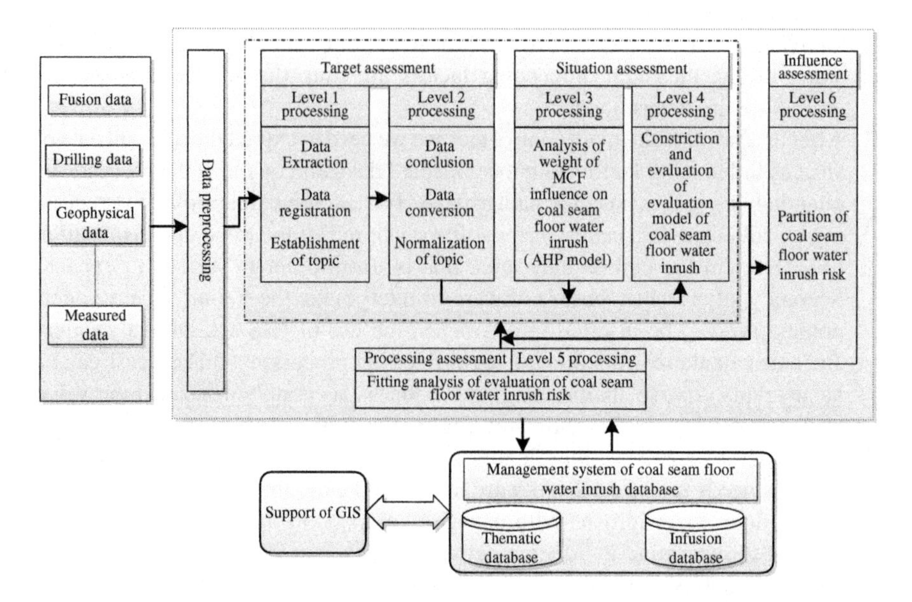

Fig. 4.7 Research flowchart

of all factors also change, the weights distribute again and are combined into a more reasonable state to service that more accurate decision. For example, when a factor indicator data is too small or too large. In order to show the effect of the factor on the comprehensive decision outcome and not be neutralized by other factors [9–18], we should appropriately increase the weight value of the factor to make comprehensive evaluation results reduced or increased compared with constant weight evaluation, which also improves the vulnerability based on constant weight.

In the past, when use the vulnerability index method to evaluate the water inrush risk of the coal seam floor, the weights of the main control factors are fixed by various mathematical methods, and the weights are fixed, and it cannot be reflected by the changes of state of main control factors. The constant weight model of constant weight cannot highlight the control of the risk of coal seam floor water inrush due to the sudden change of the main control factor value in the study area, and cannot reflect in the different combination state the relative importance and preference and their control and influence on water inrush of coal seam floor.

Therefore, this study applies the zoning method to the prediction of waterfront vulnerability of coal seam floor, which effectively improves the defects in the method of constant weight evaluation, and adjusts its weight conditions according to the change of the index value in different evaluation units and to take into account the comprehensive effect of main control factor state value in different combinations of state, effectively reflects the comprehensive effect of multiple main factors in the coal floor water inrush risk assessment, which greatly improve the accuracy of water vulnerability evaluation of floor water inrush. The evaluation results are more reasonable and accurate.

4.3.1.2 Basic Theory of Partitioned Variable Weight Model

The definition and basic principle of the partition variable weight model are as follows:

Definition 1: Constant vector. Take $W_0=(w_1^{(0)}, w_2^{(0)}, \ldots, w_m^{(0)})$ as a constant vector, if $\forall_j \in \{1, 2, \ldots, m\}$, $w_j^{(0)} \in (0, 1]^m$, and it satisfy $\sum_{j=1}^{m} w_j^{(0)} = 1$

Definition 2: Local variable weight vector Give mapping $[0, 1]^m \rightarrow (0, 1]^m$, and assume the vector $W(x)=(w_1(x), w_2(x), \ldots, w_m(x))$ as a m-dimensional partition variable weight vector, if it satisfies the condition that (1) normalization: $\sum_{j=1}^{m} w_j(x_1, x_2 \ldots x_m) = 1$;(2)Penalty incentive: $\forall_j \in \{1, 2, \ldots, m\}$ all existing α_j,$\beta_j \in [0, 1]$ and $\alpha_j \leq \beta_j$, let $w_j\,(x_1, \ldots, x_m)$ approximately x_j to monotone decrease in $\left[0, \alpha_j\right]$, but monotone increase in $\left[\beta_j, 1\right]$, If each $W_j(X)$ is continuous for all arguments, then $W(x)$ is a continuous local variable weight vector.

Definition 3: State variable weight vector Vector S: $[0, 1]^m \rightarrow (0, +\infty)^{\mathrm{m}}$, name $S(\mathrm{x}) = (S_1(\mathrm{x}), S_2(\mathrm{x}), \ldots, S_{\mathrm{m}}(\mathrm{x}))$ as a dimensioned state variable weight vector, if each $j \in \{1, 2, \ldots\ldots, m\}$, contains α_j, $\beta_j \in [0, 1]$ and $\alpha_j \leq \beta_j$, and satisfy the condition that (1) to each $j \in \{1, 2, \ldots, m\}$,and any constant weight vector $W_0=(w_1^{(0)}, w_2^{(0)}, \ldots\ldots, w_m^{(0)})$, $w_j(x)=\frac{w_j^0 \cdot S_j(X)}{\sum_{k=1}^{m} w_k^{(0)} S_k(X)}$ monotone decrease in $\left[0, \alpha_j\right]$, and monotone increase in $\left[\beta_j, 1\right]$. (2) when $0 \leq x_i \leq x_k \leq \alpha_i \wedge \alpha_k$, $S_i(x) \geq S_k(x)$; when $\beta_i \vee \beta_k \leq x_i \leq x_k \leq 1$ $S_i(X) \leq S_k(X)$ definite.

Assume $S(x)$ as a dimensioned state variable weight vector, $W_0 = (w_1^{(0)}, w_2^{(0)}, \ldots\ldots, w_m^{(0)})$ as any constant weight vector, then variable weight $W(X)$ is:

$$
\begin{aligned}
W(X) &= \frac{W_0 \cdot S(X)}{\sum_{j=1}^{m} w_j^{(0)} S_j(X)} \\
&= \left(\frac{w_1^{(0)} S_1(X)}{\sum_{j=1}^{m} w_j^{(0)} S_j(X)}, \frac{w_2^0 S_2(X)}{\sum_{j=1}^{m} w_j^0 S_j(X)}, \ldots, \frac{w_m^0 S_m(X)}{\sum_{j=1}^{m} w_j^0 S_j(X)} \right)
\end{aligned}
\tag{4.9}
$$

4.3.2 Vulnerability Evaluation Model Based on Variable Weight Theory

Based on the method of information fusion evaluation based on the constant weight, the evaluation method of coal mine floor water inrush based on partitioned variable weight theory is developed. Not only to consider a variety of main factors and their corresponding different weights, but also quantitatively determine the corresponding weight of same factor in different state values. Partition variable weight model is applied to determine the "variable weight" in different state of coal seam floor water inrush with the same main control factor, the specific steps are as follows:

4.3.2.1 Construction of Partition State Vector

The state variable weight vector is the key step to construct the variable weight evaluation model. Summed up the main control factors variable weight evaluation characteristics, according to the principle of value of the penalty having hinder effects on water inrush, and reward indicator value has promoting effects on water inrush, to construct the partition variable weight vector in line with the objective laws of the water inrush [33–36]. Combined with the above analysis, the mathematical formula that determine the state variable weight vector of the same factor is as follows: Eq. 4.10, the same factor state variable weight vector curve is shown in Fig. 4.8.

$$S_j(x) = \begin{cases} e^{a_1(d_{j1}-x)} + c - 1, & x \in [0, d_{j1}) \\ c, & x \in [d_1, d_{j2}) \\ e^{a_2(x-d_{j2})} + c - 1, & x \in [d_{j2}, d_{j3}) \\ e^{a_3(x-d_{j3})} + e^{a_2(d_{j3}-d_{j2})} + c - 1, & x \in [d_{j3}, 1] \end{cases} \tag{4.10}$$

Among which c, a_1, a_2, a_3 are adjusting weight vector, d_{j1}, d_{j2}, d_{j3} are the Jth factor variable interval threshold.

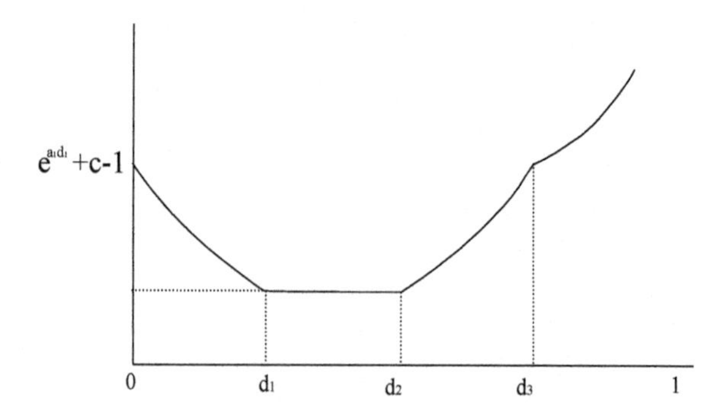

Fig. 4.8 State variable weight vector of same factor

The state variable weight vector is an exponential vector. The larger the value of a in the state variable vector, the greater the penalty for the factor, that is, the greater the obstruction of the water inrush, once the factor state value is less than the interval threshold d_{j1}, When the factor state value is greater than the interval threshold d_{j2}, the greater the incentive, the more the promotion of the occurrence of water inrush. For the adjustment of the level of C, generally the C is smaller, the greater the degree of punishment and incentives, in opposite, it will be more obscure. The threshold of variable weight interval d_{j1}, d_{j2}, d_{j3} divide the same factor into four intervals:

- [0, d_{j1})belongs to the punishment interval;
- [dj1, dj2) belongs to the constant excitation interval;
- [dj2, dj3) belongs to the initial stimulate interval;
- [dj3, dj4) belongs to the initial excitation interval;
- [Dj3,1] belongs to the strong excitation interval.

4.3.2.2 Determination of the Threshold of the State Variable Vector Interval

When applying the variable weight model to evaluate the water inrush vulnerability of the coal seam floor, it is necessary to determine the variable range of the main control factors first. The variable range determines the type of weight adjustment on the main control factor 's index in study area. According to a certain mechanism and the different state values, the state value of each factor is divided into four areas: "initial incentive", "strong incentive", "punishment" and "constant" to get the aim of partition variable weight. However, the determination of variable threshold is now a difficult point, and there is no unified analyzing determined method. In this paper, there are some differences in the spatial distribution of the main control factors in the bottom of the coal seam, and there is a certain similarity. In this paper, there are some differences in the spatial distribution of the main control factors in the floor of the coal seam, and there is a certain similarity. A method of determining the threshold of the variable range is proposed: K-mean value clustering method [19–26].

Its core idea is to randomly select K data points as the initial cluster center in the data set, and then calculate the distance of the concentrated data points to the center of each initial cluster. According to the shortest distance principle, the data points are divided into clusters closest to the center of the initial cluster. And then repeat the above algorithm, respectively, calculate the new center of each cluster. So, give iteration like this, until the new center is not changing, then it illustrates clustering algorithm convergence. Otherwise it needs to re-determine the initial cluster center, re-divide all data points. Until the requirements are met, the K-mean value clustering algorithm is shown in Fig. 4.9.

The K-means method is used to classify the main control factor data. According to the needs of the state variable vector, it divides the classification into four categories. And according to the classified results to determine the various main control factors index's classification threshold ($f_{j1}, f_{j2}, f_{j3}, f_{j4}, f_{j5}, f_{j6}$). And, according to the following calculation formula to determine the various factors variable range threshold.

$$d_{j1} = \left(f_{j1} + f_{j2}\right)/2 \tag{4.11}$$

$$d_{j2} = \left(f_{j3} + f_{j4}\right)/2 \tag{4.12}$$

$$d_{j3} = \left(f_{j5} + f_{j6}\right)/2 \tag{4.13}$$

where: d_j is the variable range threshold;
f_j is the classification threshold of index value;
j is the factor in cluster grading

4.3.2.3 Determination of the Weighting Parameters of the State Variable Weight Vector

The key step in constructing the state variable weight vector is to determine the parameters of its weighting. These parameters can control and adjust the weighting

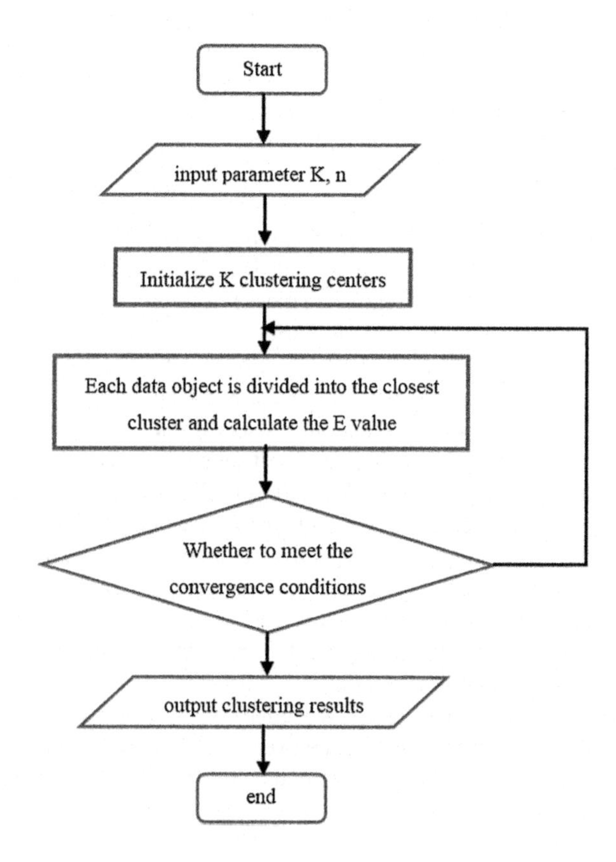

Fig. 4.9 K-means algorithm flowcharts

effect of weight, and play the corresponding "penalty" and "incentive" role. However, due to the study of variable weight theory is not perfect, for the determination of weighting parameters the current related research is still very seldom [37, 38]. In this paper, a method to determine the weighting parameters in the coal seam floor water inrush evaluation model that can realize expected weighting parameters. The technical steps are as follows:

[1] Select or give an evaluation unit that meets the constraints. Under the premise that the variable range interval threshold is obtained, the selected evaluation unit should satisfy the following constraint conditions: factor state values respectively are $x_1, x_2, x_3, \ldots, x_n$, among which x_1 is in the punishment interval,x_2 is in no punishment on motivation interval, x_3 is in initial motivation interval, x_4 is in strong motivation interval. There is no limitation in other intervals of factors state value, at the same time **n** vectors constant weight weighted value $(w_1^0, w_2^0, w_3^0 \ldots w_n^0)$ is known.

[2] Determine the ideal weight weighted value of the selected evaluation unit. The determined method can comprehensively consider the role of each factor index value and determine it by consulting with the relevant experts, but it can be determined also according to the attitude of decision-makers. The constant weight weighted value $(w_1, w_2, w_3 \ldots w_n)$ of the selected evaluation unit is determined, and the ideal variable weight weighted value $(w_1, w_2, w_3 \ldots w_n)$ of various main control factor that satisfy the decision preference of decision-makers.

[3] According to the following parameter formula to solve the transfer parameter value.

$$w_1 = \frac{w_1^0\left[e^{a_1(d_{11}-x_1)} + c - 1\right]}{\sum_{j=1}^{n} w_j^0 s_j(x)} \tag{4.14}$$

$$w_2 = \frac{w_2^0 c}{\sum_{j=1}^{n} w_j^0 s_j(x)} \tag{4.15}$$

$$w_3 = \frac{w_3^0\left[e^{a_2(x_3-d_{32})} + c - 1\right]}{\sum_{j=1}^{n} w_j^0 s_j(x)} \tag{4.16}$$

$$w_4 = \frac{w_4^0\left[e^{a_3(x_4-d_{43})} + e^{a_2(d_{43}-d_{42})} + c - 2\right]}{\sum_{j=1}^{n} w_j^0 s_j(x)} \tag{4.17}$$

$$a_1 = \frac{1}{(d_{11}-x_1)} \mathrm{In}\left[\frac{w_1 w_2^0 - w_2 w_1^0}{w_2 w_1^0} c + 1\right] \tag{4.18}$$

$$a_2 = \frac{1}{(x_3-d_{32})} \mathrm{In}\left[\frac{w_3 w_2^0 - w_2 w_3^0}{w_2 w_3^0} c + 1\right] \tag{4.19}$$

$$a_3 = \frac{1}{(x_4 - d_{43})} \ln\left[\frac{w_4 w_2^0 - w_2 w_4^0}{w_2 w_4^0} c + 2 - \left(\frac{w_3 w_2^0 - w_2 w_3^0}{w_2 w_3^0} c + 1\right)^{\frac{(d_{43}-d_{42})}{(x_3-d_{32})}}\right] \tag{4.20}$$

$$k_1 c = (k_2 c + 1)^{k_3} - 1 \tag{4.21}$$

Where: $k_1 = \frac{w_2^0 - w_2^0(w_1+w_2+w_3+w_4) - w_2\left(w_5^0+w_6^0+w_7^0\right)}{w_2 w_5^0}$;

$$k_2 = \frac{w_1 w_2^0 - w_2 w_1^0}{w_2 w_1^0};$$

$$k_3 = \frac{d_{51} - x_5}{d_1^1 - x_1};$$

$x_1, x_2, x_3, \ldots x_n$ are factor index value;
$d_{11}, d_{12}, d_{13}, d_{21}, d_{22}, d_{23}, \ldots d_{n1}, d_{n2}, d_{n3}$ are threshold of variable interval;
$w_1^{(0)}, w_2^{(0)}, w_3^{(0)} \ldots w_n^{(0)}$ are weighted value of factor constant weight;
$w_1, w_2, w_3, \ldots w_n$ are weighted value of factor variable weight.

4.3.2.4 Determination of the Weighting Weights of the Main Control Factors

The variable weights of the main control factors are determined by using the partitioned variable weight model. The mathematical expression is as follows:

$$W(X) \triangleq \frac{W_0 \cdot S(X)}{\sum_{j=1}^m w_j^{(0)} S_j(X)} \triangleq \left(\frac{w_1^{(0)} S_1(X)}{\sum_{j=1}^m w_j^{(0)} S_j(X)}, \frac{w_2^0 S_2(X)}{\sum_{j=1}^m w_j^0 S_j(X)}, \ldots, \frac{w_m^0 S_m(X)}{\sum_{j=1}^m w_j^0 S_j(X)}\right) \tag{4.22}$$

where: $S(X)$—m is Dimensional state variable weight vector
$W_0 = (W_1^{(0)}, W_2^{(0)}, \ldots\ldots, W_m^{(0)})$—any constant vector;
$W(X)$—m Dimensional variable weight vector.

4.3.2.5 Establishment of Vulnerability Evaluation Model Based on Partitioned Variable Weight Theory

Based on constructing main control factor system that influence water inrush in coal seam, the vulnerability index method based on partitioned variable weight theory is used, we can not only consider the various main factors mentioned above and their corresponding weights, but also quantitatively determine the same factors corresponding weights in different status values. The vulnerability index model of coal seam floor is as follows:

$$VI = \sum_{i=1}^{m} w_i(Z) \cdot f_i(x, y) = \sum_{i=1}^{m} \frac{w_i^{(0)} S_i(Z)}{\sum_{j=1}^{M} w_i^{(0)} S_i(Z)} f_i(x, y)$$

$$\sum_{j=1}^{m} w_j(Z) = 1 \tag{4.23}$$

where: VI—vulnerability index;
W_i—influencing factors variable weight vector;
$f_i(x, y)$—single factor influencing value function;
(x, y)—geographical coordinates;
$w^{(0)}$—any constant weight vector;
$S(Z)$—m dimension partition state variable weight vector.

4.4 Chapter Summary

This paper first uses the information fusion technology to study the multi-source information fusion architecture and fusion algorithm of water inrush evaluation, and establishes the JDL-based water inrush evaluation information fusion model based on GIS. Using AHP and ANN technology, this paper gives a complete set of determining the weight of the main control method. Based on the above two points of study, the author summarizes the vulnerability evaluation method based on the theory of constant weight. Based on the vulnerability evaluation method based on the theory of constant weight, the vulnerability index method based on partition variable weight is studied. The method of constructing the state vector of each main control factor is analyzed, and the method of determining the variable range (threshold), transfer parameter and variable weight is summarized.

References

1. Wang, Jizhou. 2011. *Study on Reinforcement of Coal Seam Floor Under High Temperature and High Pressure*. Handan: Hebei University of Engineering (in Chinese).
2. I. Bloch. 1996. Information combination operators for data fusion: A comparative review with classification. *IEEE Transactions on Systems, Man and Cybernetics Part A* 26 (1): 52–67.
3. Robin, R. 1988. Murphy. Dempster-Shafer theory for sensor fusion in autonomous mobile robots. *IEEE Transactions on Robotics and Automation* 14 (2): 197–206.
4. P.M. Atkinson, Tatnall, A.R.L. 1997. Neural Networks in Remote Sensing. *International Journal of Remote Sensing* 18 (4): 699–709.
5. Han, Liqun. 2005. *Theory, Design And Application of Artificial Neural Network*. Beijing: Chemical Industry Press (in Chinese).
6. Wang, Wanliang. 2008. *Artificial Intelligence and Its Application*. Beijing: Higher Education Press (in Chinese).
7. Kanaya, F., and S. Miyaker. 1992. Bayes statistical behavior and valid generalization of pattern classifying neural networks. *IEEE Transaction on Neural Networks* 3: 471–475.

8. Li Zhong, Shunian Ning, Jinde Zhang et al. Study on ANN model of mine environment evaluation. *Computer Engineering and Applications* 43 (16): 238–240 (in Chinese).
9. Zhang, Yingping. 2012. *Application of Variable Weight Model for Quality Evaluation of Urban Engineering Geological Environment*. Bei Jing: Chinese Academy of Geological Sciences (in Chinese).
10. Shuqiao, Duan. 2003. A method of variable weight synthetic evaluation for safety management of power business. *Mathematics in Practice and Theory* 33 (8): 17–23.
11. Fan, Maofei, and Guohua Chen. 2008. Study on risk assessment method of gas fired power plant based on variable weight model. *Fuzzy Systems and Mathematics* 28 (4): B48–B53 (in Chinese).
12. Wang, Peizhuang. 1985. *Fuzzy Set and Shadow of Random Set*. Beijing: Beijing Normal University press (in Chinese).
13. Hongxing, Li. 1995. Factors spaces and mathematical frame of knowledge representation(VIII)—Variable weights analysis. *Fuzzy Systems and Mathematics* 9 (3): 1–9. (in Chinese).
14. Li, Hongxing. 1996. Factor sapces and mathematical frame of knowledge representation (IX) structure of balance functions and Weber—Fechner's characteristics. *Fuzzy Systems and Mathematics* 10 (3): 12–17 (in Chinese).
15. Liu, Wenqi. 1998. The penalty-incentive utility in variable weight synthesizing. *Systems Engineering-Theory & Practice* 18 (4): 41–47 (in Chinese).
16. Liu, Wenqi. 1998. The impulsing strategy and its algorithm of variable weight synthesizing. *Systems Engineering-Theory & Practice* 18 (12): 40–43 (in Chinese).
17. Liu, Wenqi. 2000. The Ordinary Variable Weight Principleand Multiobjective Decision-Making. *Systems Engineering-Theory & Practice* (3): 1–11 (in Chinese).
18. Yao, Bingxue, and Hongxing Li. 2000. Axiomatic System of Local Variable Weight. *Systems Engineering-Theory & Practice* 20 (1): 106–109 (in Chinese).
19. Li Dexing, and Hongxing Li. 2002. The properties and construction of state variable weight vectors. *Journal of Beijing Normal University (Natural Science)* 38 (4): 455–461 (in Chinese).
20. Li Dexing, and Hongxing Li. 2004. Analysis of variable weights effect and selection of appropriate state variable weights vector in decision making. *Control and Decision* 19 (11): 1241–1245 (in Chinese).
21. Weights transferring effect of state variable weights vector. *Systems Engineering-Theory & Practice*, 2009, 29 (6): 127–131 (in Chinese).
22. Cui, Hongmei, Yundong Gu, and Kuiming Sun, et al. 2004. Some notes on the system of axioms for state variable weights and variable weights. *Journal of Beijing Normal University (Natural Science)* 40 (1): 1–7 (in Chinese).
23. Hou, Haijun, Gu Yundong, and Wang Jiayin. 2005. State variable weight vectors construction based on functions. *Fuzzy Systems and Mathematics* 19 (4): 119–124 (in Chinese).
24. Xu. Zezhong. 2010. Construction of variable weight vectors used in variable weight decision-making. *Journal of Liaoning Technical University (Natural Science)* 29 (5): 843–846 (in Chinese).
25. Axiomatic System of State Variable Weights and Construction of Balance Functions. *Systems Engineering-Theory & Practice*, 1999, 19 (7): 116–118 (in Chinese).
26. Li, Yueqiu. 2008. *Variable weight synthesis theory and multi objective decision making*. Kunming: Kunming University of Science and Technology (in Chinese).
27. Wang, M.W., P. Xu, J. Li, et al. 2014. Anovel set pair analysis method based on variable weights for liquefaction evaluation. *Natural Hazards* 70 (2): 1527–1534.
28. Xu, Shubo. 1998. *Principles of analytic hierarchy process*. Tianjin: Tianjin University press (in Chinese).
29. Wu, Qiang, Zhenquan Jiang, and Yunlong, Li. 2003. Study on ground fissure disaster in Shanxi fault basin. Beijing: Geological Publishing House (in Chinese).
30. Zhang, Jincai, Yuzhuo Zhang, Tianquan Liuet al. 1997. Seepage of rock mass and water inrush from coal seam floor. Beijing: Geological Publishing House (in Chinese).

31. Wu Qiang, Fan Zhenli, Liu Shouqiang et al. 2011. Water-richness evaluation method of water-filled aquifer based on the principle of information fusion with GIS: Water-richness index method. *Journal of China Coal Society* 36 (1): 1124–1128 (in Chinese).
32. Liya, Zhang, and Li Deqing. 2009. An ideal point approach of determining state variable weights vector in decision making. *Mathematics in Practice and Theory* 39 (6): 93–97. (in Chinese).
33. Wu, Qiang, Bo Li, and Shouqiang Liu et al. 2013. Vulnerability assessment of coal floor groundwater bursting based on zoning variable weight model: A case study in the typical mining region of Kailuan. *Journal of China Coal Society* 09: 1516–1521 (in Chinese).
34. Taur, J.S., and S.Y. Kung. 1993. Fuzzy decision neural network and application to data fusion interference and neural network. *Information Sciences* 71 (1): 27–41.
35. Agterberg F.P., and G.F. Bonham-Carter. 1990. Weights of evidence modeling: A new approach to mapping mineral potential statistical application in the Earth science. *Geology Survey of Canada* 89 (9): 171–183.
36. G.F. Bonhamr Carter, and F.P. Agterberg. 1990. *Weights of Evidence: A New Approach to Mapping Mineral Potential, Statistical Applications in the Earth Sciences*, 231–245. Canada: Geological Survey of Canada.
37. Ohlmacher Gregory, C., and John C. Davis. 2003. Using multiple logistic regression and GIS technology to predict landslide hazard in Northeast Kansas, USA. *Engineering Geology* 69: 331–343.
38. T.L. Saaty. 1980. *The Analytic Hierarchy Process*. New York: McGraw-Hill.

Chapter 5
Analysis of Vulnerability Index Method Based on Variable Weight Theory in Engineering Application

5.1 Application of ANN Vulnerability Index Method Based on Variable Weight—A Case Study of Xiandewang Mine with "Monoclinic Type"

Xiandewang Mine is located at the junction of Shahe City, Hebei Province and Wu'an City. This mine belongs to the Baiquan spring area system, and the groundwater flow in the spring area system is mainly from west to east, and the tendency of the coal stratum in the Xiandewang Mine is the same as flow direction of the coal field, and the flow field superposition relation of coal-bearing stratum and karst water is a typical "monoclinic order type". The study takes the example of risk evaluation of water inrush in Ordovician limestone aquifer in #9 of this coal mine, and the weight of the main control factor is determined by ANN, and then introduces the variable weight model. The dynamic weight of each main control factor is determined by establishing the variable weight vector. And finally applies variable weight theory to partition the assessment of the vulnerability of water inrush in Ordovician limestone in the 9# coal floor of Xiandewang Mine.

5.1.1 General Situation of Mine Area

The eastern boundary of the Xiandewang Mine is a fault; the northern part is bounded by K2 and K3 exploration lines; the western boundary is the outcropping line of coal seam. The southern boundary is a mixed boundary of artificial boundary and fault. Mining area is basically piedmont platform, and the terrain changes greatly. The study area has four distinctive seasons, its climate type is temperate continental climate. There are only three streams belonging to the tributaries of the Beiming River, and the streams are affected by the rainfall, and show that have water in them in rainy season and cutoff in dry season. The average annual relative humidity is 66.0%. Annual average wind speed is approximately 3 m/s. In addition, the mining area is in seismic intensity zoning of 6–7 M, according to historical records that

Y. Zeng, *Research on Risk Evaluation Methods of Groundwater Bursting from Aquifers Underlying Coal Seams and Applications to Coalfields of North China*, Springer Theses, https://doi.org/10.1007/978-3-319-79029-9_5

the surrounding counties once occurred earthquake of 6–7.5 M, so there is still the possibility of seismic activity in this area.

5.1.2 Geological and Hydrogeological Backgrounds

5.1.2.1 Geological Background

(1) **Formation**

The study area is covered by the Quaternary loose sedimentary layer. According to the previous geological data, the strata developed in this area are Ordovician, Carboniferous, Permian and Quaternary from bottom to top. The Carboniferous Taiyuan formation and the Permian Shanxi Formation are coal-bearing stratum in the region. The area has a total of 19 coal bearing strata, and the thickness of the coal seam reaches to 13.54 m, among which 8 are available or local available coal seams.

(2) **Geological structure**

The structures developed in the area where Xiandewang coalfield is located are mainly Longyao South normal fault, Taihang mountain major dislocation and Xiandewang normal fault (Fig. 5.1). The extension length of Longyao South normal fault is approximately 44 km in the field, showing a direction of EW. The study area is located on the upper plate (south side) where is the general development area of coal-bearing stratum in Longyao south normal fault; the Taihang mountain major dislocation is extended in the NNE direction, and the drop is approximately between 500 and 1800 m. The Xiandewang Coal Mine that was studied is located in the east side of the dislocation. The normal fault of Xiandewang is presented in the field in the form of F1 fault, the fault extends along the NNE direction, leans to westward and the drop is in the range from 50 to 100 m, at the same time this fault is also a boundary of Xiandewang Coal Mine and the adjacent Zhangcun Coal Mine.

As shown in the mine field structure outlines (Fig. 5.2), the faults in the mining area are mainly NNE-direction faults, and the folds in the area are mainly eastern NNE-direction Xiandewang syncline, southern NWW-direction Luanxie syncline and western NW-direction Li Shigang syncline. The form of anticline is not obvious, but the form syncline is more obvious and cut by normal fault. From the extending direction of the folds and fault, it can be seen that the main stress in the area is from north, west and east.

5.1.2.2 Hydrogeological Background

According to the control law of the groundwater flow field caused by tectonic cutting and the groundwater flow field of the Ordovician limestone fissure aquifer, the Xingtai mining area can be divided into Baiquan hydrogeological unit, Shiguquan

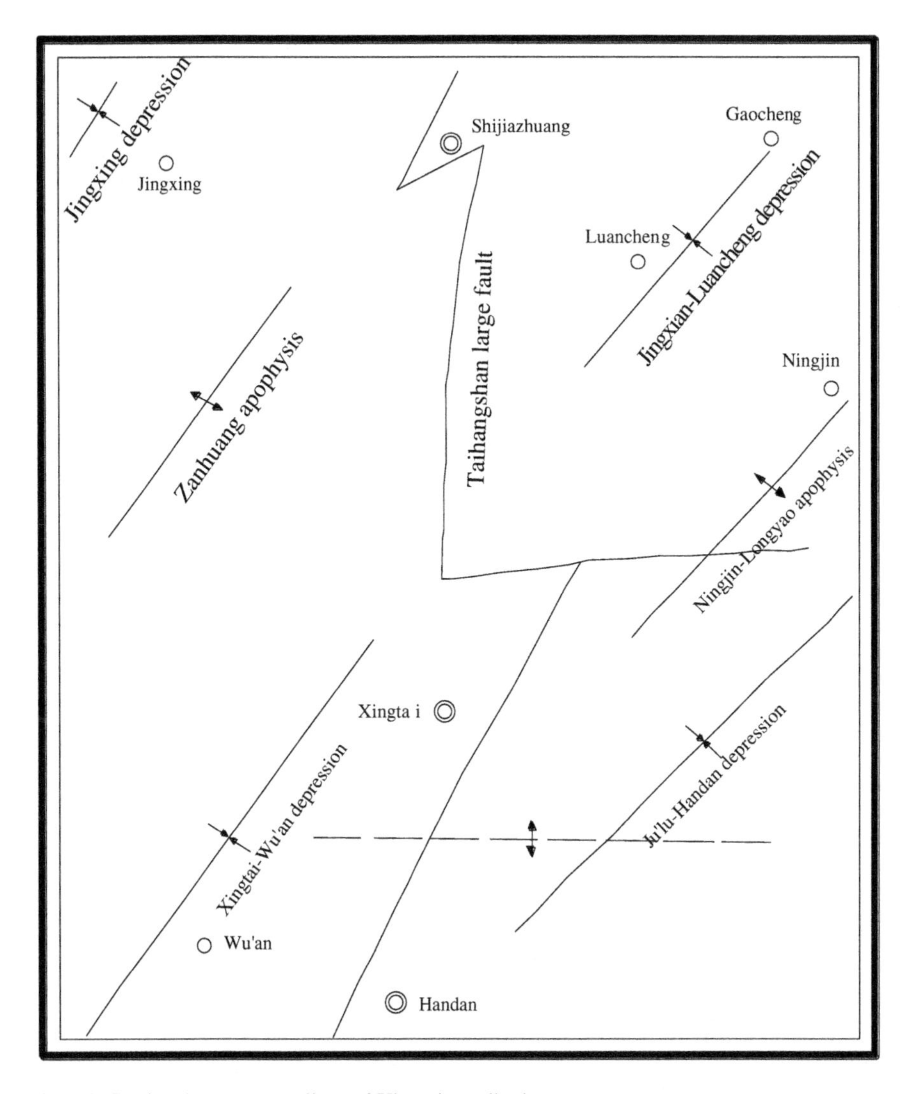

Fig. 5.1 Regional structure outlines of Xingtai ore district

hydrogeological unit and Longyao hydrogeological unit, the study area is located in the Baiquan hydrogeological unit. Xiandewang mining area is all covered by the Cenozoic strata, in the north of it is the coal seam hidden outcrop area. F1 fault at the east side of the field and does not completely cut off the hydraulic linkages between the aquifers on both sides, so the east part should be weakly permeable boundary; the west side has exposed Ordovician aquifers, so the Ordovician aquifer can not only be supplied by to the surface water system, but also in the rainy season can accept the supply of atmospheric precipitation; the south side of the north side are

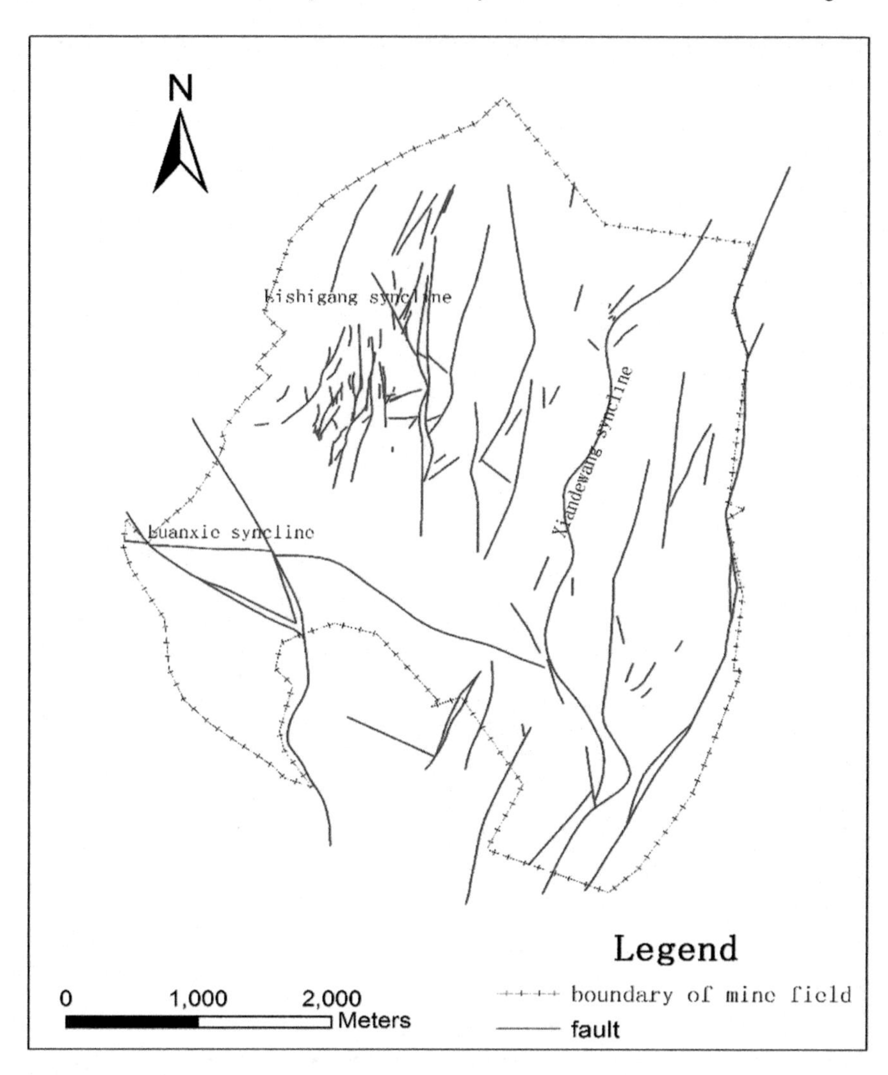

Fig. 5.2 Structural outlines of Xiandewang ore district

the natural water boundary, which are impacted by human' activities like the daily mine production.

In the study area, there are Ordovician limestone aquifer, Taiyuan limestone aquifer, Permian sandstone aquifer, and Quaternary pore aquifer. The relationship between the aquifers and coal seam's location is shown in Fig. 5.3.

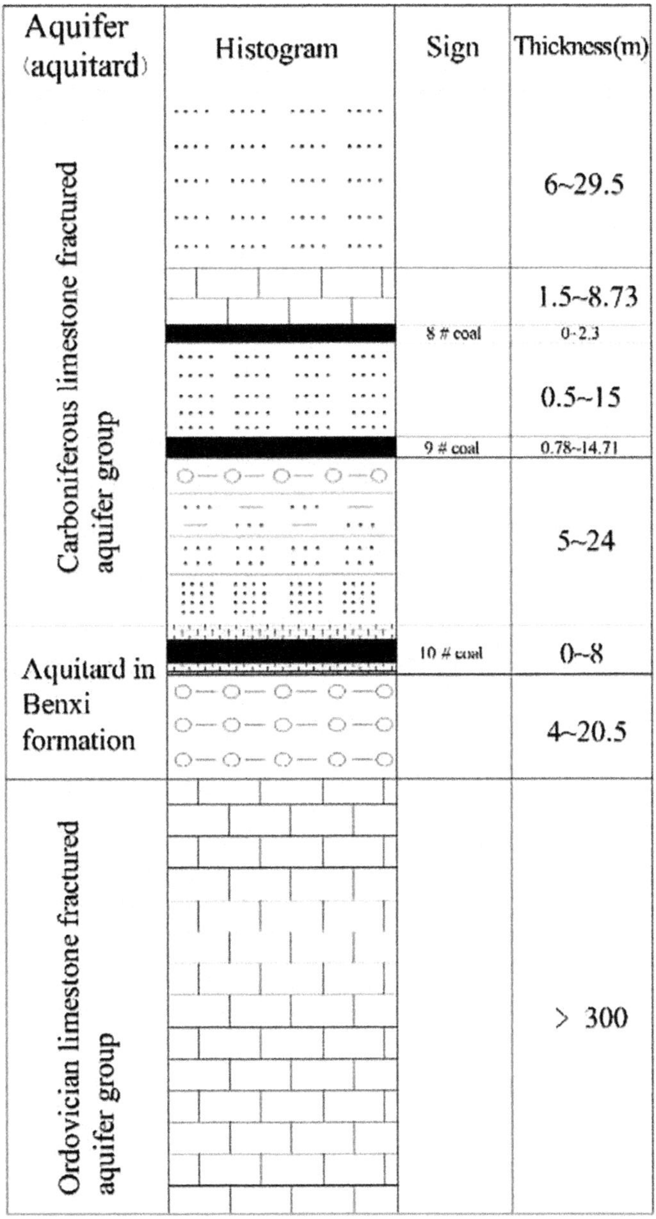

Aquifer (aquitard)	Histogram	Sign	Thickness(m)
Carboniferous limestone fractured aquifer group			6~29.5
			1.5~8.73
		8 # coal	0-2.3
			0.5~15
		9 # coal	0.78~14.71
			5~24
Aquitard in Benxi formation		10 # coal	0~8
			4~20.5
Ordovician limestone fractured aquifer group			> 300

Fig. 5.3 Generalized columnar section of $9^{\#}$ seam's floor and aquifer

5.1.3 Analysis and Determination of Pressure Area

Comparison of the relationship between the standard level of the Ordovician limestone water and the stand level of the #9 coal floor confirmed that the #9 coal mine is going to explore the pressure area of the Ordovician limestone. The distribution of the Ordovician limestone pressure area is shown in Fig. 5.4.

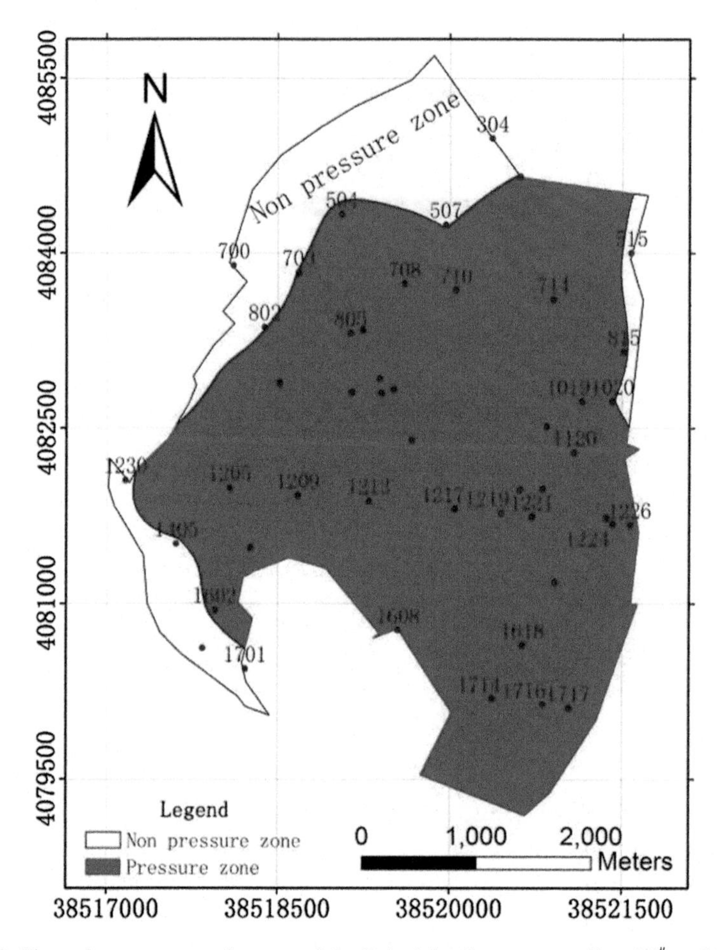

Fig. 5.4 The under-pressure zoning map of the Ordovician limestone aquifer of $9^{\#}$ coal floor

5.1.4 Study on Main Controlling Factors of Ordovician Limestone Water Inrush in #9 Coal Floor

Based on the study of hydrogeological conditions, mine hydrogeological conditions and geological conditions of Xiandewang Coal Mine, the main factors influencing the water inrush of #9 coal floor in the study area are analyzed mainly from the aspects of floor aquifer, aquiclude and tectonic condition. Nine main control factors are determined that influence #9 coal mine floor water inrush. The nine factors are the aquifer water abundance, the water pressure of the aquifer, the equivalent thickness of the effective aquifers, the thickness of brittle rock in the mine pressure damage zone, the water resistance of the ancient weathering crust, the distribution of the fault and the fold axis, the distribution of intersection of the fault and the fold axis, fault scale index and distribution of collapse column.

5.1.5 Establishment of Thematic Map of Main Controlling Factor of Floor Water Inrush

After determining the main control factors, the first need to do is extract the data of the main control factors in the collected data, and use Surfer's interpolation function and GIS spatial analysis function to quantify the data, and then establish the main control factor thematic Layer (Fig. 5.5). Because of the existence of the ancient weathering crust in the Ordovician limestone top interface, so this study does not believe that there is Ordovician confined water raising belt; for the calculation of the depth of the mining pressure damage zone, due to the study area has no measured data, so this paper took empirical formula to calculate; there are many faults in the study area and they not evenly distributed. Therefore, the fault is divided into 100 m $\times$ 100 m unit meshes in the area where the fault is developed. In the area where the fault is not developed, divide it according to the 500 m $\times$ 500 m element mesh.

5.1.6 Vulnerable Evaluation of Water Inrush of #9 Coal Floor Based on ANN's Constant Weight Model

5.1.6.1 Artificial Neural Network (ANN) Model to Determine the Weight

Because of the non-linear relationship between the water inrush of the coal seam floor and the factors influencing the water inrush, the study uses the ANN learning training and the reasoning and inducing function to obtain the size of weight of that influencing factors that influence #9 floor water inrush. The network structure of commonly used BP model in ANN is shown in Fig. 5.6.

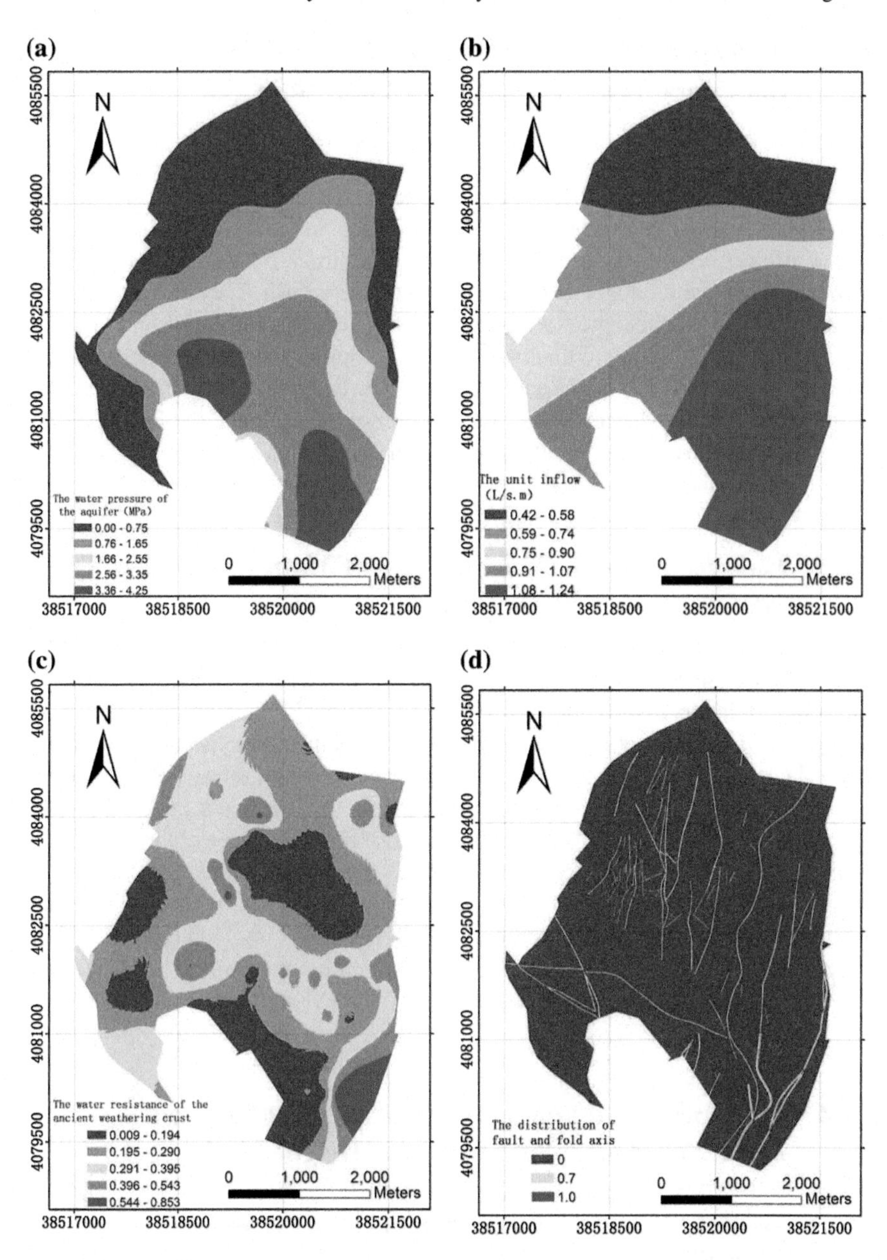

Fig. 5.5 Thematic layer graph of factors that influence #9 coal seam floor water inrush [**a** the water pressure of the ordovician limestone aquifer; **b** the water abundance of the ordovician limestone aquifer; **c** the water resistance of the ancient weathering crust; **d** the distribution of fault and fold axis; **e** the distribution of the endpoints and intersections of the fault; **f** the thickness of the brittle rock under the floor mine pressure damage zone; **g** the scale index of the fault; **h** the equivalent thickness of the effective aquifer; **i** distribution of collapse column]

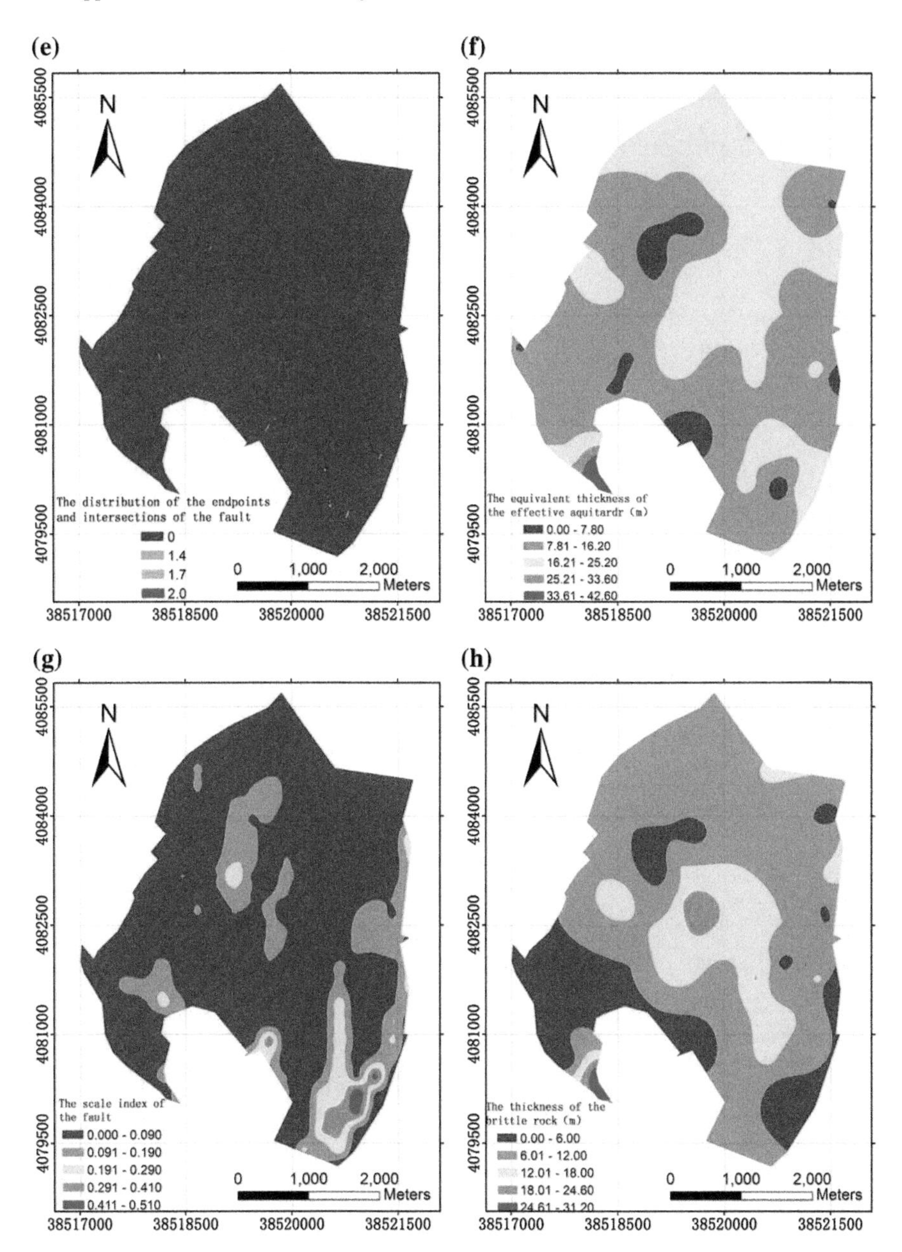

Fig. 5.5 (continued)

According to the construction principle of BP neural network and the selection experience of each parameter, the main parameters of BP network constructed in this paper are shown in Table 5.1.

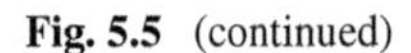

Fig. 5.5 (continued)

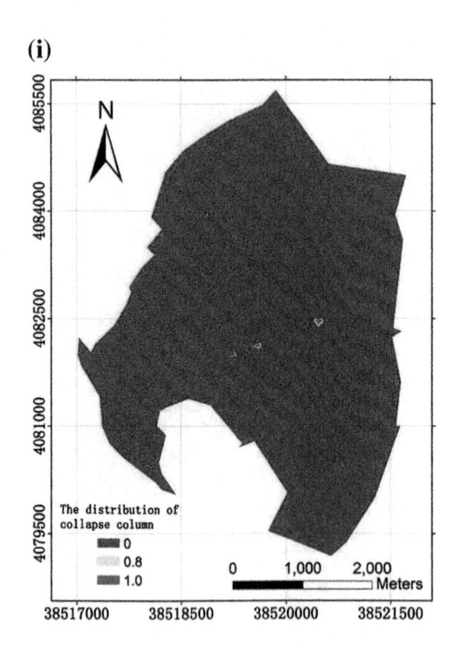

Table 5.1 Parameters of BP artificial neural network

Parameter type	Number of neurons in input layers	Number of hidden layers	The number of neurons in hidden layers	The number of neurons in output layers
Number	9	1	19	1

Table 5.2 Weights of different influencing factors that influence coal floor seam's water inrush

Influencing index	(W_1)	(W_2)	(W_3)	(W_4)	(W_5)	(W_6)	(W_7)	(W_8)	(W_9)
Weight Wi	0.2132	0.2020	0.0478	0.1235	0.0396	0.1283	0.0516	0.0415	0.1525

Note (W_1)—weight of Ordovician limestone aquifer pressure; (W_2)—weight of Ordovician limestone water abundance; (W_3)—weight of water resistance of ancient weathering crust; (W_4)—weight of distribution of fault and fold axis; (W_5)—weight of end points and intersection distribution in fault; (W_6)—weight of equivalent thickness of effective aquifer; (W_7)—weight of scale index of fault; (W_8)—weight of brittle rock thickness under the mine pressure damage zone (W_9)—weight of collapse column distribution

According to the BP neural network parameters, it is necessary to select the samples to conduct the learning training on the established network to get the relationship between the neural network neurons, but this is not the relationship between main control factor and the final evaluation results. So, the relationship between the neurons should be analyzed and processed, and use the related coefficient, the related index and the absolute influence index to obtain the contribution of the main control factors to the final evaluation result. That is weight of the control factor (Table 5.2).

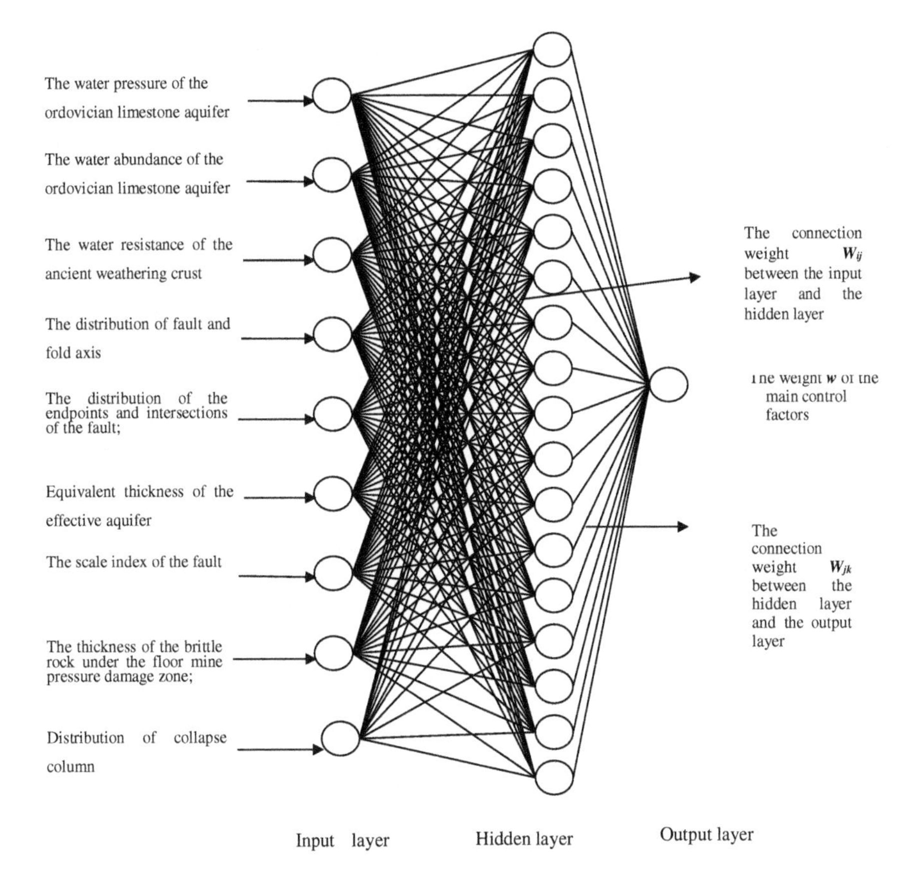

Fig. 5.6 Structure diagram of BP artificial neural network

5.1.6.2 Evaluation of Water Inrush Vulnerability of #9 Coal Floor Based on ANN Constant Weight Model

(1) **Construction of thematic layer of data normalization and the master control normalization**

Due to the different dimensions of the main control factors, the dimensionality of the main control factors should be eliminated before the vulnerability evaluation, so that the different main control factors are comparable. Take a normalized approach to eliminate the dimension, due to the main control factors and the floor water inrush have the different relativity. Some main control factors are positive correlation factors, such as the water pressure of the Ordovician aquifer, the water abundance of the Ordovician aquifer layer, and the distribution of fault and fold axis. And the distribution of fault and fold axis, etc., And some influencing factors are negatively correlated with the water inrush floor, such as the water resistance of the ancient weathering crust, the equivalent thickness of the effective aquifers, and the brittle rock thickness beneath the mine pressure damage zone. Therefore, in the normalization process, we should first determine the correlation between the main control factor and the floor water inrush. For the positive correlation factor, the maximum value method is used to normalize, and the negative correlation factor is normalized by the minimum method, For the distribution of the collapse column, the distribution of the fault and the fold axis, the end point of the fault and the fold axis are normalized by the eigenvalue method.

(2) **Establishment of vulnerability evaluation model and information fusion**

According to the weight determined by ANN, the vulnerability index method evaluation model was established:

$$
\begin{aligned}
VI = \sum_{k=1}^{n} W_k \cdot f_k(x, y) &= 0.2132\mathrm{f}_1\,(\mathrm{x, y}) + 0.2020\mathrm{f}_2\,(\mathrm{x, y}) + 0.0478\mathrm{f}_3\,(\mathrm{x, y}) \\
&+ 0.1235\mathrm{f}_4\,(\mathrm{x, y}) + 0.0396\mathrm{f}_5\,(\mathrm{x, y}) + 0.1283\mathrm{f}_6\,(\mathrm{x, y}) + 0.0516\mathrm{f}_7\,(\mathrm{x, y}) \\
&+ 0.0415\mathrm{f}_8\,(\mathrm{x, y}) + 0.1525\mathrm{f}_9\,(\mathrm{x, y})
\end{aligned}
$$

where: *VI* is the vulnerability index;
W_k is the weight of main control factor;
$f_k\,(x, y)$ is single factor influencing index function;
x and y are geographic coordinates, n is the number of influencing factor.

Then we use GIS to analyze and process the dimensionless layers, and then form new topological relations according to different attributes, to further determine the number of units of vulnerability index.

(3) **Evaluation and division of vulnerability of water inrush in coal seam floor**

Natural Breaks were used to analyze the vulnerability index of each unit, finally the thresholds of the vulnerability assessment of #9 coal seam floor were determined as follows: 0.26, 0.36, 0.45, 0.54 (Fig. 5.7). The vulnerability index has a positive correlation with the risk of water inrush, and its numerical size can directly reflect the risk of water inrush. Finally, the risk of water inrush in the study area was divided into five grades according to the threshold:

- $VI > 0.54$ Coal seam floor water inrush vulnerable zone
- $0.45 < VI \le 0.54$ Coal seam floor water inrush more vulnerable zone
- $0.36 < VI \le 0.45$ Coal seam floor water inrush transition zone
- $0.26 < VI \le 0.36$ Coal seam floor water bursts safer zone
- $VI < 0.26$ Coal seam floor water inrush relative safe zone.

After determining the threshold in the partition, we can obtain the evaluation partition of the Ordovician limestone water inrush vulnerability of the #9 coal floor of the Xiandewang Coal Mine based on the constant weight model (Fig. 5.8). It can be

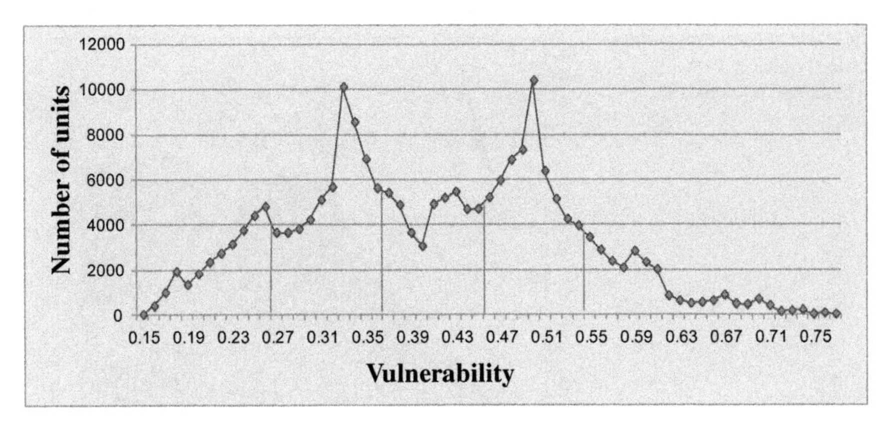

Fig. 5.7 Frequency histogram of vulnerable index of mine floor

seen from the figure that the risk assessment of the water inrush in the Ordovician limestone of the #9 coal mine in the Xiandewang Mine can be divided into five grades. In general, the risk of water inrush in the northern part of the study area is small, and the risk of water inrush in the southern and southeastern part of the study area is high.

5.1.7 Vulnerable Evaluation of Water Inrush of #9 Coal Floor Based on Partition Variable Weight Model

(1) **Construction of partition state variable weight vector and determination of adjust weight parameter**

According to the evaluation characteristics of coal seam floor water inrush, the variable weight model can reflect the influence of the change of the main control factor value on the evaluation result more accurately. According to the positive and negative correlation between the index value and the floor water inrush, the "penalty" and "incentive" measures were taken to further highlight or weaken the impact on the floor water inrush. Since the contribution of different indicators on the floor water inrush is different, the state value is divided into four intervals, namely, the strong excitation interval [d3,1], the initial excitation interval [d2, d3), the non-penalty non-excitation interval [d1, D2), punishment interval [0, d1). The state variable weight vector constructed in this paper not only takes into account the excitation effect of the state value which plays a role in promoting the water inrush from the floor, but also consider the effect of the state value which has the obstructing effect on the water inrush. The state vector curve is shown in Fig. 5.9.

The mathematical model of state variable vector is determined as follows:

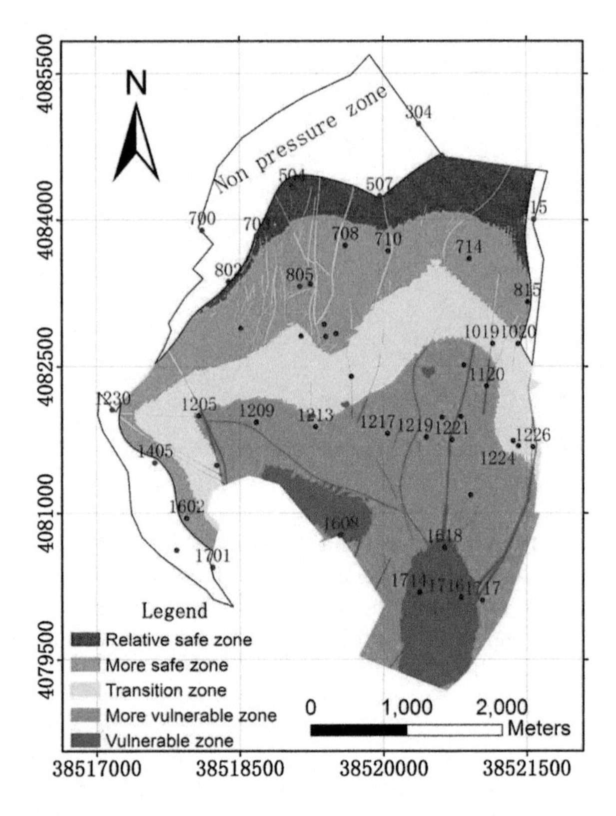

Fig. 5.8 Division evaluation of the vulnerable index method base on constant weight

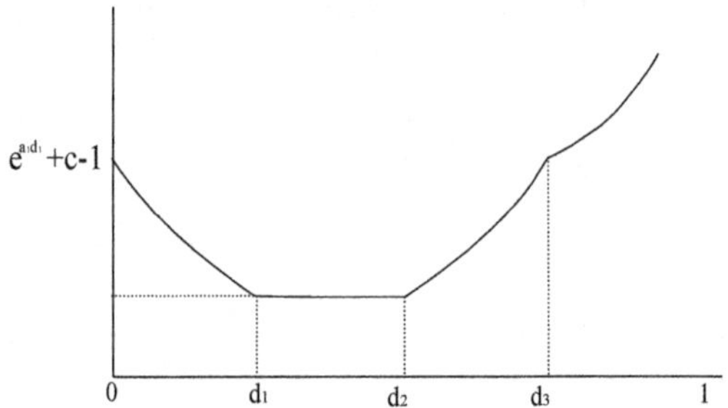

Fig. 5.9 State variable weight vector of different main controlling factors

$$S_j(x) = \begin{cases} e^{a_1(d_1-x)} + c - 1, & x \in [0, d_1) \\ c, & x \in [d_1, d_2) \\ e^{a_2(x-d_2)} + c - 1, & x \in [d_2, d_3) \\ e^{a_3(x-d_3)} + e^{a_2(d_3-d_2)} + c - 2, & x \in [d_3, 1) \end{cases} \tag{5.1}$$

where a1, a2, a3, c are the weighting parameters of the state variable weight vector; d1, d2, d3 are the variable range thresholds of the main control factors. After repeated analysis and debugging, when the ideal variable effect is achieved, the values of the weighting parameters of the state variable weight vector are respectively a1 = 0.241, a2 = 0.197, a3 = 0.611, c = 0.073.

(2) **Determination of main control factors variable weight range**

In this evaluation, the K-means clustering method in dynamic clustering is used to analyze the factor index value. The clustering method selects the iterative classification according to the K-means algorithm. According to the construction of the variable weight model, the classification category is 4 categories, And, then the eight factors like water pressure of aquiclude, the thickness of the aquifers which affect the floor water inrush are divided respectively, the index value classification threshold of main control factors are achieved, after obtaining the average the threshold in variable range can be obtained. For the two main control factors of fault distribution, distribution of intersection of the fault and the fold. Their index value is fixed, and the value is small. In this evaluation, the threshold of the two factors is determined by combining the previous application experience. And finally get the corresponding main control variable weight range (Table 5.3).

(3) **Establishment of variable weight model and determination of weight of variable weight**

Establish partition variable weight model (Formula 5.2) of 9# coal seam floor water inrush main control factors, then use this model to calculate the corresponding weights (Table 5.4) when the index values of main control factors are different.

$$W(X) \triangleq \frac{W_0 \cdot S(X)}{\sum_{j=1}^{9} w_j^{(0)} S_j(X)} \triangleq \left(\frac{w_1^{(0)} S_1(X)}{\sum_{j=1}^{9} w_j^{(0)} S_j(X)}, \frac{w_2^0 S_2(X)}{\sum_{j=1}^{9} w_j^0 S_j(X)}, \ldots, \frac{w_9^0 S_9(X)}{\sum_{j=1}^{9} w_j^0 S_j(X)} \right) \tag{5.2}$$

(4) **Establishment of vulnerability index method for partitioned variable weight model**

Based on the analysis of hydrogeological conditions and water inrush factors, this paper according to the various weight of various main control factors determined by partition variable weight model, establishes the evaluation model of water inrush vulnerability of #9 coal seam in Xiandewang mining area. The model formula is as follows:

Table 5.3 Variable weight intervals of each main controlling factor of #9 coal seam

Main control factors in variable weight range	Punishment interval	Non-penalty non-excitation interval	Initial excitation interval	Strong excitation interval
Water pressure of aquiclude	$0.257 > x \geq 0$	$0.538 > x \geq 0.257$	$0.822 > x \geq 0.538$	$1 \geq x \geq 0.822$
Water abundance of aquiclude	$0.294 > x \geq 0$	$0.522 > x \geq 0.294$	$0.778 > x \geq 0.522$	$1 \geq x \geq 0.778$
Water resistance of ancient weathering crust	$0.161 > x \geq 0$	$0.505 > x \geq 0.161$	$0.764 > x \geq 0.505$	$1 \geq x \geq 0.764$
Distribution of fault and fold axis		$0.500 > x \geq 0$	$0.800 > x \geq 0.500$	$1 \geq x \geq 0.800$
Distribution of endpoints and intersection of faults and fold axis		$0.350 > x \geq 0$	$0.500 > x \geq 0.350$	$1 \geq x \geq 0.500$
Equivalent thickness of effective aquiclude	$0.278 > x \geq 0$	$0.572 > x \geq 0.278$	$0.743 > x \geq 0.572$	$1 \geq x \geq 0.743$
Scale index of fault	$0.231 > x \geq 0$	$0.475 > x \geq 0.231$	$0.709 > x \geq 0.475$	$1 \geq x \geq 0.709$
brittle rock thickness under mining pressure damage zone	$0.315 > x \geq 0$	$0.590 > x \geq 0.315$	$0.742 > x \geq 0.590$	$1 \geq x \geq 0.742$
Collapse column		$0.500 > x \geq 0$	$0.800 > x \geq 0.500$	$1 \geq x \geq 0.800$

Table 5.4 Variable weights of main controlling factors

Water pressure variable weight	Water abundance variable weight	Ancient weathering crust variable weight	Structure distribution variable weight	Fault end point and intersection distribution variable weight	Thickness of the effective aquifer variable weight	Fault index variable weight	Brittle rock thickness under damage zone various weight	Distribution of collapse column various weight
0.203	0.168	0.075	0.15	0.082	0.13	0.05	0.039	0.103
0.234	0.219	0.062	0.144	0.036	0.131	0.038	0.039	0.097
0.247	0.208	0.062	0.148	0.051	0.129	0.041	0.032	0.082
0.221	0.187	0.056	0.141	0.033	0.168	0.038	0.058	0.098
0.224	0.169	0.057	0.131	0.08	0.156	0.037	0.053	0.093
0.195	0.194	0.054	0.218	0.03	0.137	0.038	0.043	0.091
0.184	0.156	0.065	0.224	0.074	0.122	0.039	0.042	0.094
0.195	0.168	0.077	0.146	0.087	0.141	0.042	0.041	0.103
0.164	0.241	0.069	0.188	0.068	0.119	0.033	0.039	0.079
0.161	0.153	0.082	0.247	0.071	0.117	0.037	0.038	0.094
0.183	0.173	0.088	0.149	0.088	0.133	0.04	0.041	0.104
0.183	0.18	0.099	0.116	0.092	0.138	0.038	0.053	0.101
0.236	0.226	0.057	0.138	0.039	0.129	0.041	0.053	0.081
0.23	0.217	0.064	0.142	0.033	0.132	0.038	0.044	0.099
0.227	0.198	0.055	0.208	0.03	0.116	0.039	0.039	0.088
0.241	0.171	0.064	0.196	0.031	0.113	0.039	0.036	0.109
0.247	0.171	0.077	0.147	0.034	0.137	0.049	0.044	0.103
0.15	0.348	0.053	0.123	0.052	0.13	0.033	0.046	0.065
0.167	0.157	0.061	0.225	0.08	0.14	0.037	0.039	0.095
0.256	0.22	0.054	0.139	0.032	0.124	0.037	0.039	0.099

(continued)

Table 5.4 (continued)

Water pressure variable weight	Water abundance variable weight	Ancient weathering crust variable weight	Structure distribution variable weight	Fault end point and intersection distribution variable weight	Thickness of the effective aquifer variable weight	Fault index variable weight	Brittle rock thickness under damage zone various weight	Distribution of collapse column various weight
0.155	0.314	0.05	0.095	0.075	0.141	0.034	0.047	0.088
0.233	0.185	0.054	0.142	0.037	0.158	0.037	0.056	0.098
0.169	0.24	0.068	0.188	0.067	0.119	0.031	0.039	0.079
0.223	0.181	0.068	0.156	0.036	0.142	0.042	0.043	0.109
0.19	0.179	0.069	0.116	0.091	0.16	0.042	0.045	0.108
0.185	0.171	0.064	0.151	0.088	0.156	0.046	0.036	0.103
0.253	0.199	0.055	0.14	0.032	0.141	0.038	0.045	0.097
0.253	0.199	0.056	0.14	0.03	0.14	0.037	0.048	0.097
0.155	0.313	0.05	0.095	0.071	0.141	0.038	0.046	0.091
0.231	0.183	0.053	0.141	0.037	0.163	0.034	0.059	0.099
0.223	0.173	0.058	0.136	0.068	0.148	0.047	0.048	0.099
0.237	0.217	0.068	0.142	0.033	0.137	0.031	0.048	0.087
0.236	0.219	0.062	0.141	0.033	0.131	0.038	0.043	0.098
0.225	0.219	0.068	0.141	0.033	0.131	0.038	0.047	0.098
0.245	0.259	0.069	0.158	0.029	0.149	0.042	0.037	0.012
0.244	0.229	0.06	0.138	0.032	0.126	0.033	0.042	0.096
0.228	0.186	0.057	0.141	0.033	0.162	0.038	0.058	0.098
0.242	0.224	0.058	0.139	0.034	0.122	0.037	0.045	0.099
0.182	0.26	0.074	0.122	0.072	0.129	0.033	0.042	0.085

(continued)

Table 5.4 (continued)

Water pressure variable weight	Water abundance variable weight	Ancient weathering crust variable weight	Structure distribution variable weight	Fault end point and intersection distribution variable weight	Thickness of the effective aquifer variable weight	Fault index variable weight	Brittle rock thickness under damage zone various weight	Distribution of collapse column various weight
0.158	0.315	0.051	0.095	0.072	0.141	0.033	0.047	0.088
0.221	0.186	0.057	0.141	0.033	0.168	0.038	0.058	0.098
0.226	0.201	0.049	0.208	0.03	0.135	0.025	0.035	0.091
0.227	0.189	0.055	0.213	0.029	0.115	0.034	0.039	0.099
0.223	0.21	0.051	0.211	0.029	0.115	0.035	0.039	0.087
0.229	0.187	0.057	0.141	0.033	0.162	0.038	0.055	0.098
0.198	0.237	0.065	0.122	0.077	0.129	0.041	0.042	0.089
0.23	0.187	0.057	0.142	0.033	0.158	0.038	0.056	0.099
0.188	0.323	0.066	0.097	0.03	0.129	0.037	0.039	0.091
0.224	0.182	0.053	0.143	0.034	0.169	0.048	0.048	0.099
0.229	0.187	0.057	0.141	0.033	0.162	0.038	0.055	0.098
0.243	0.168	0.076	0.147	0.034	0.137	0.046	0.046	0.103
0.215	0.191	0.052	0.197	0.087	0.108	0.032	0.036	0.083
0.183	0.159	0.045	0.203	0.079	0.161	0.038	0.047	0.085
0.224	0.185	0.056	0.143	0.032	0.166	0.038	0.058	0.098
0.211	0.177	0.053	0.133	0.079	0.173	0.036	0.055	0.083
0.218	0.138	0.048	0.21	0.053	0.164	0.033	0.052	0.084
0.227	0.172	0.057	0.136	0.08	0.148	0.037	0.048	0.095
0.183	0.156	0.068	0.224	0.08	0.122	0.036	0.037	0.094
0.171	0.157	0.057	0.225	0.08	0.14	0.037	0.038	0.095

(continued)

Table 5.4 (continued)

Water pressure variable weight	Water abundance variable weight	Ancient weathering crust variable weight	Structure distribution variable weight	Fault end point and intersection distribution variable weight	Thickness of the effective aquifer variable weight	Fault index variable weight	Brittle rock thickness under damage zone various weight	Distribution of collapse column various weight
0.151	0.305	0.048	0.123	0.072	0.137	0.033	0.046	0.085
0.249	0.221	0.06	0.137	0.032	0.125	0.037	0.042	0.096
0.237	0.185	0.046	0.141	0.033	0.165	0.038	0.057	0.098
0.231	0.206	0.052	0.21	0.026	0.112	0.037	0.038	0.088
0.231	0.187	0.055	0.142	0.033	0.156	0.038	0.059	0.099
0.227	0.185	0.056	0.14	0.033	0.167	0.038	0.058	0.098
0.243	0.199	0.065	0.14	0.032	0.141	0.038	0.045	0.097
0.222	0.2	0.056	0.213	0.03	0.117	0.035	0.038	0.089
0.228	0.206	0.059	0.179	0.039	0.115	0.037	0.039	0.098
0.228	0.186	0.056	0.141	0.033	0.162	0.038	0.058	0.098
0.219	0.151	0.052	0.206	0.072	0.136	0.031	0.044	0.089
0.244	0.213	0.057	0.136	0.042	0.126	0.037	0.048	0.097
0.228	0.188	0.057	0.141	0.033	0.159	0.038	0.061	0.095
0.208	0.169	0.051	0.213	0.03	0.152	0.035	0.053	0.089
0.234	0.171	0.035	0.216	0.029	0.148	0.035	0.051	0.081
0.252	0.22	0.056	0.139	0.032	0.124	0.038	0.042	0.097
0.228	0.176	0.058	0.141	0.038	0.159	0.038	0.065	0.097
0.192	0.164	0.069	0.235	0.031	0.129	0.035	0.046	0.099
0.204	0.159	0.076	0.229	0.032	0.128	0.037	0.039	0.096

(continued)

Table 5.4 (continued)

Water pressure variable weight	Water abundance variable weight	Ancient weathering crust variable weight	Structure distribution variable weight	Fault end point and intersection distribution variable weight	Thickness of the effective aquifer variable weight	Fault index variable weight	Brittle rock thickness under damage zone various weight	Distribution of collapse column various weight
0.183	0.173	0.067	0.149	0.088	0.154	0.04	0.043	0.104
0.212	0.198	0.055	0.21	0.029	0.145	0.034	0.039	0.078
0.231	0.203	0.049	0.213	0.029	0.112	0.036	0.041	0.086
0.209	0.167	0.051	0.226	0.03	0.141	0.035	0.051	0.09
0.219	0.146	0.052	0.201	0.073	0.138	0.038	0.046	0.087
0.229	0.204	0.054	0.21	0.029	0.113	0.034	0.038	0.088
0.242	0.221	0.059	0.145	0.032	0.126	0.037	0.042	0.096
0.212	0.181	0.056	0.14	0.032	0.184	0.038	0.06	0.097
0.218	0.178	0.043	0.26	0.028	0.113	0.034	0.038	0.088
0.187	0.157	0.07	0.227	0.08	0.124	0.032	0.038	0.085
0.202	0.171	0.052	0.214	0.03	0.153	0.035	0.053	0.09
0.227	0.185	0.056	0.14	0.033	0.167	0.038	0.058	0.098
0.237	0.219	0.061	0.139	0.031	0.135	0.036	0.044	0.098
0.236	0.221	0.063	0.141	0.038	0.128	0.032	0.043	0.098

Note Because of the large amount of data, there are only part data selected here

$$
\begin{aligned}
VI = \sum_{i=1}^{9} W_i(z) \cdot f_i(x, y) = \sum_{i=1}^{9} \frac{w_i^{(0)} S_i(X)}{\sum_{j=1}^{9} w_j^{(0)} S_j(X)} f_i(x, y) = \frac{w_1^{(0)} S_1(X)}{\sum_{j=1}^{9} w_j^{(0)} S_j(X)} f_1(x, y) \\
+ \frac{w_2^{(0)} S_2(X)}{\sum_{j=1}^{9} w_j^{(0)} S_j(X)} f_2(x, y) + \frac{w_3^{(0)} S_3(X)}{\sum_{j=1}^{9} w_j^{(0)} S_j(X)} f_3(x, y) \\
+ \frac{w_4^{(0)} S_4(X)}{\sum_{j=1}^{9} w_j^{(0)} S_j(X)} f_4(x, y) + \frac{w_5^{(0)} S_5(X)}{\sum_{j=1}^{9} w_j^{(0)} S_j(X)} f_5(x, y) \\
+ \frac{w_6^{(0)} S_6(X)}{\sum_{j=1}^{9} w_j^{(0)} S_j(X)} f_6(x, y) + \frac{w_7^{(0)} S_7(X)}{\sum_{j=1}^{9} w_j^{(0)} S_j(X)} f_7(x, y) \\
+ \frac{w_8^{(0)} S_8(X)}{\sum_{j=1}^{9} w_j^{(0)} S_j(X)} f_8(x, y) + \frac{w_9^{(0)} S_9(X)}{\sum_{j=1}^{9} w_j^{(0)} S_j(X)} f_9(x, y)
\end{aligned}
$$

where: *VI*—Vulnerability index;
Wi—Influencing factors variable weight vector;
W(0)—Any constant weight vector;
fi (x, y)—The i single factor influencing value function;
(x, y)—Geographical coordinates;
$S(x) = \{S_1(0), S_2(x), \ldots S_9(x)\}$—9-dimensional partition state variable weight vector.

(5) **Vulnerability evaluation partition of coal seam floor water inrush based on the partitioned variable weight model**

The vulnerability index (VI) of each stacking unit in the study area is calculated by the above formula. The greater the value, the more likely the occurrence of water inrush. And then use the natural breaks (Jenks) commonly used on the grading map to classify the vulnerability index value, and then determine the threshold of the partition 0.25, 0.39, 0.55, 0.67, and get the #9 coal seam floor Ordovician limestone water inrush vulnerability index method evaluation partition map based on the partition weight theory (Fig. 5.10).

- VI > 0.67 Coal seam floor water inrush vulnerable zone
- 0.55 < VI ≤ 0.67 Coal seam floor water inrush more vulnerable zone
- 0.39 < VI ≤ 0.55 Coal seam floor water inrush transition zone
- 0.25 < VI ≤ 0.39 Coal seam floor water bursts less safe zone
- VI ≤ 0.25 Coal seam floor water inrush relative safe zone.

(6) **Identification and testing of model**

Only part of the working face of #9 coal seam have a formal backstoping, there is no actual water point data to fit, this paper selects the four drills 703,1714, meso, #9 water source well as fitting point (Fig. 5.11).

Through the testing analysis, the selected fitting point of the location are consistent with the theory, so this evaluation results of model are ideal.

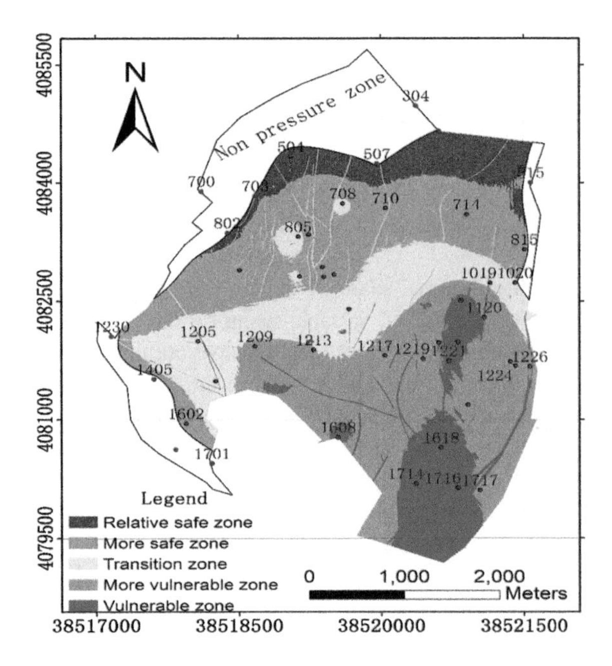

Fig. 5.10 Water inrush vulnerability evaluation zoning map of $9^{\#}$ coal seam floor to Ordovician limestone aquifer base on variable weight

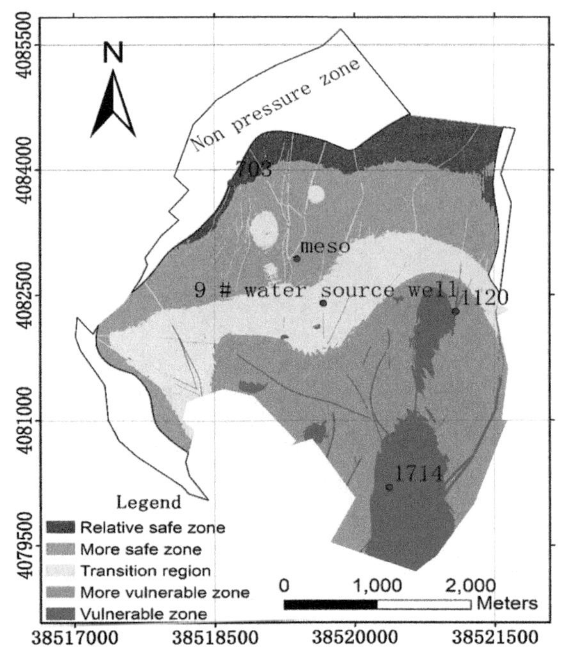

Fig. 5.11 Water inrush vulnerability evaluation zoning fitting map of $9^{\#}$ coal seam floor to Ordovician limestone aquifer base on variable weight

5.1.8 Comparative Analysis of Variable Weight Model, Constant Weight Model, and Water Inrush Coefficient Method

5.1.8.1 Contrastive Analysis of Evaluation Results of Variable Weight Model and Constant Weight Model

Compare the coal seam floor water inrush vulnerability evaluation partition result graph (Fig. 5.12) of variable weight model and traditional constant weight model, and combine the two models to have comparative analysis, the results of the two evaluations are matched in most of the lot. The possibility of water inrush from the northwest to the southeast of the study area is gradually increasing, and the vulnerable areas are mainly concentrated in the south of the study area. However, it is obvious that there are significant differences in the evaluation results in local areas, local difference region (A, B, C) are selected to be compared.

Combine two models to analyze, we can see that the variable weight model can better reflect the influence of index value of the sudden change of main control factors on the evaluation result than the traditional constant weight model. If an index of one region happens sudden changes, for the positive impact factors, the effect of the variable effect is "incentive"; for the reverse factor, variable effect is reflected in the "punishment." For example, in the A area in above figure, if the thickness of effective aquifer gets thinned, and the area is where the water abundance increased,

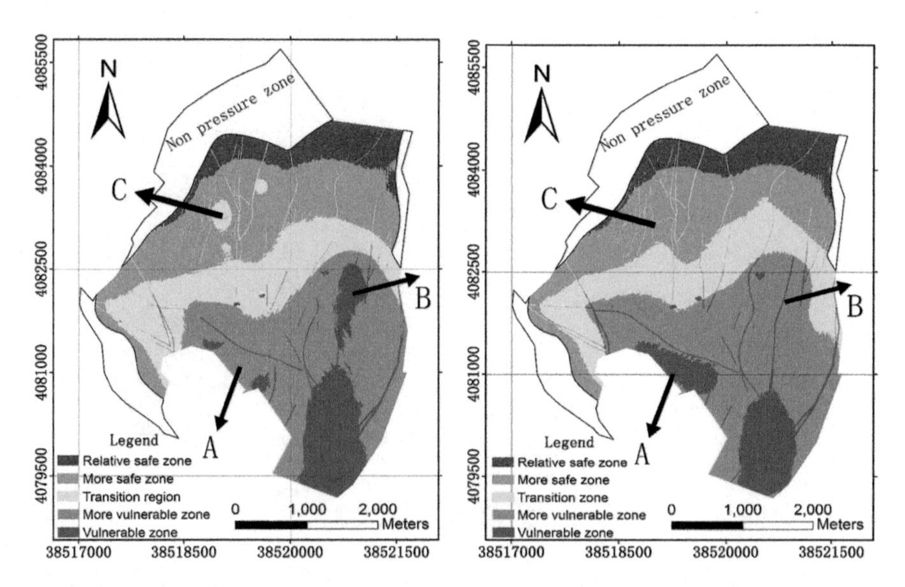

Fig. 5.12 Vulnerability evaluation comparison graph based on variable weight model and constant weight model (The left picture shows the variable weight model evaluation partition, whereas the right picture is the constant weight model evaluation partition.)

the variable weight model will give a certain degree of "incentive" to the area where the change of the index value is found. It shows in that the orange more vulnerable region in the evaluation results of constant weight model changes into red vulnerable region in the evaluation results of variable weight model. As in the B area above, due to the development of fault and pleat axis, the variable weight model gives a certain "incentive" to the weight of this main control factor, it shows like A area that the orange more vulnerable region in the evaluation results of constant weight model changes into red vulnerable region in the evaluation results of variable weight model. In C area, because of the tectonic development and the increase of the water pressure over the surrounding areas, the "incentive" to the weights from the variable weight model is represented that the more save area in evaluation results of constant weight model changed into transition region in the evaluation results of variable weight model. The variable weight model can not only consider the corresponding importance of each factor in the weighing of the process of floor water inrush, but also can effectively consider the control effect of sudden change of state value of factors in unit on floor water inrush. The more important thing is it can take into account the controlling effects of indicators with multiple factors in different combinations on water inrush, which effectively avoids the indicators that have controlling effects being neutralized by other indicators.

5.1.8.2 Comparative Analysis of the Evaluation Results of Variable Weight Model and Water Inrush Coefficient Method

In addition, the traditional water inrush coefficient method is used to evaluate the risk of Ordovician limestone water inrush in 9# coal seam. Due to the tectonic development of the study area, the critical water inrush coefficient is 0.06 MPa/m, and finally divide the water inrush risk with the threshold of water inrush coefficient = 0.06. Comparison of the evaluation results of variable weight model and the water inrush coefficient method (Fig. 5.13).

Based on the analysis of the two, we can see that water inrush coefficient method only divides $9^{\#}$ coal seam water inrush risk into two areas: safe area and danger area, and there is no buffer area. This result can not completely reflect that actual situation and cannot take advantage of guidance in the actual production in coal mine. The vulnerable index method evaluation partition map of coal seam floor Ordovician limestone water inrush based on the partition variable weight model comprehensively consider the multi- factors' influence, and consider the vulnerable index value obtained from syntagmatic relations of the index value of various factors. According to the natural classification method to divide the study area into five grades in accordance to the size of the risk of water inrush, this kind of results is closer to the actual situation.

Comparatively analyze from hydrogeological evaluation formula: the evaluation of water inrush coefficient method is only the conclusion that get from considering the thickness the #9 coal seam floor and the water head pressure undertaken by the bottom of the aquifer. Between the two is only a ratio without the concept of

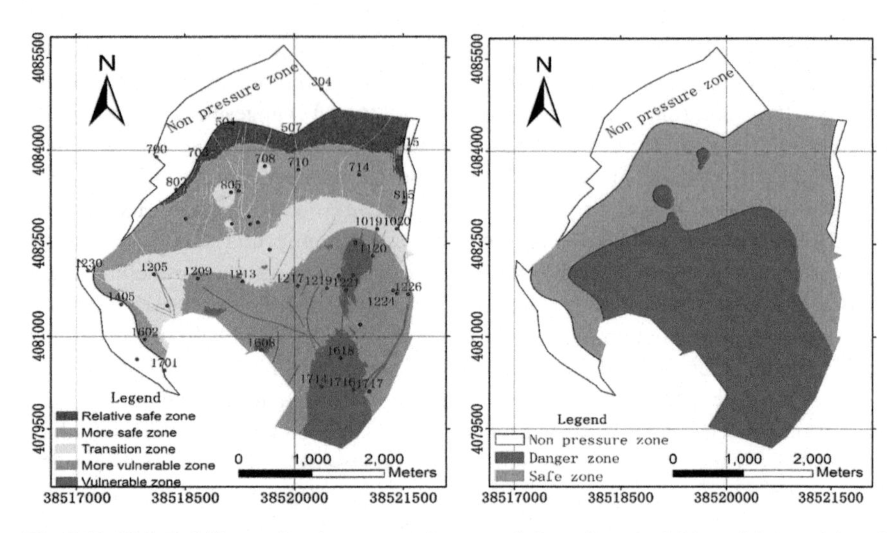

Fig. 5.13 Vulnerability evaluation comparison graph based on variable weight model and water inrush coefficient method (The left picture shows the variable weight model evaluation partition, whereas the right picture is the evaluation partition of water inrush coefficient method.)

"weight". The evaluation of the vulnerability of coal seam floor under the variable weight model is based on the comprehensive consideration of the nine main control factors from the three aspects of aquiclude geological structure and aquifer, and the variable weight model is used to reflect the influence of sudden change of factor index on the risk of water inrush. In comparison the variable weight involves more factors, and its evaluation results are more in line with the actual situation.

5.2 Application of AHP-Type Vulnerability Index Method Based on Variable Weight—A Case Study of Wangjialing Mine with "Monoclinic Inversion Type"

On March 28, 2010, Shanxi Wangjialing coal mine occurred a flooding accident with 153 people trapped in the mine which cause international attention. After the 8 days and 8 nights of the rescue, 115 people were successfully saved, but 38 people died. Wangjialing Coal Mine is in the southwest of Shanxi Province, located in the southern tip of the Loess Plateau, west of the Yellow River. Its general location is in the lower reaches of the middle valley area of the Yellow River. The shape of the whole study area is long and narrow with the direction of southwest-northeast. The mining area is approximately 119.7 km^2. Because of the small interval between the 10# coal and the Ordovician aquifer in the mine and the tectonic development of the local area, therefore, accurately evaluating the threat of Ordovician aquifer on 10$^{\#}$ coal is very important to guide the safely and efficiently exploit #10. The Wangjialing mine belongs to the Yuyuankou spring system, and the Ordovician limestone water flows from the northeast to the southwest, but the tendency of the coal-bearing strata is opposite to flow direction. The superposition of the coal-bearing strata and the karst water flow field is a typical monoclinic inversion type. In this study, take the example of evaluation on risk of Ordovician limestone aquiclude water inrush. AHP was used to determine the weight of each main control factor, and then the variable weight model was introduced. By establishing the state variable weight vector to determine the corresponding value of each factor. And finally applies partition variable weight theory to evaluate the water—inrush vulnerability of the Ordovician limestone in the 10# coal seam of Wangjialing Mine.

5.2.1 General Situation of Mine Area

Wangjialing coal mine is located in the southwest of Xiangning County, Shanxi Province, in southern part it shares the boundary with Hejin County, the western boundary of the mine is closer to the Yellow River, and opposite the Yellow River is the Hancheng mining area in Shaanxi Province whose administrative division is under the Xiangning County. The line of Zaoling Xiangnanling, Yangjiage Duo, Shangxi village to the west of the coal mine, and it east to the line of the east of western Jiaokouciangling, Aoding village; north to the line of Changning village, Zhangma, LIugeyuan village; and south to the line of Qianan in Zaolingxiang, Gujian, Lingshang village. The scope of the study is shown in Fig. 5.14.

Because the mine is located on the Loess Plateau, its special area leads to the obvious change in the climate of this region. The climate type is semi-arid continental monsoon climate. The precipitation in the study area is mainly concentrated in the third quarter, the precipitation time distribution is very uneven, the average annual precipitation is 567.2 mm for many years, and the maximum and minimum annual precipitation are 767.4 mm and 385.4 mm respectively. The maximum annual evaporation can reach 2346.4 mm, the frozen depth of the soil can reach 61 cm. The surface water in the field is the Yellow River Basin, E River and Shunyi River. The surface water inside the mine belongs to reaches of Yellow River, E river, Shunyi river drainage. Yellow River flows from north to south outside the western boundary of the field, and inside the area is seasonal river. Shunyi River flows from northeast to southwest through the middle-west part of the field, and E river is flows from east to west through the northeast part of this field.

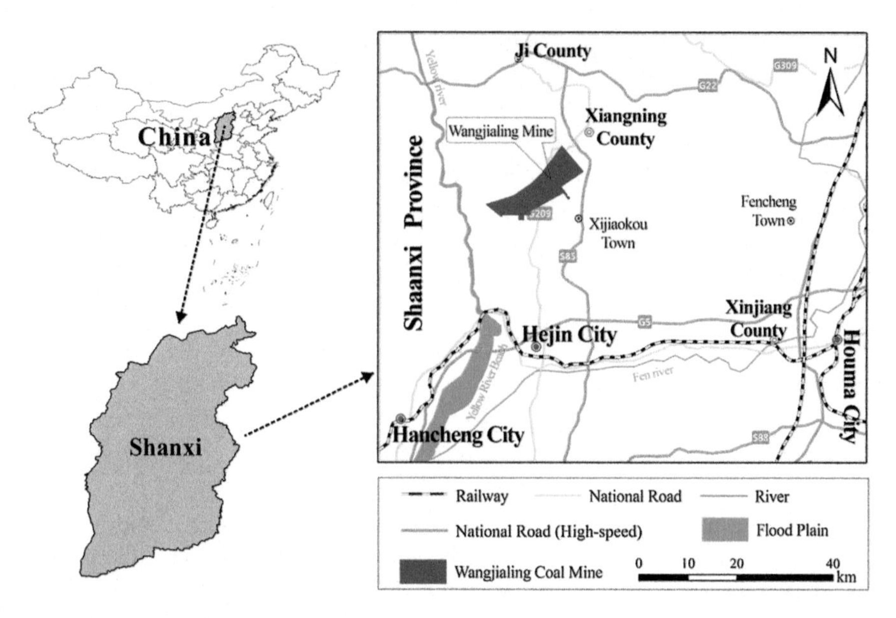

Fig. 5.14 Location of Wangjialing mine

5.2.2 Geological and Hydrogeological Backgrounds

5.2.2.1 Geological Backgrounds

(1) **Lithology**

The Quaternary loess is developed in most of the field, but in a small number of areas, due to the influence of the valley, the upper and lower layers of the Permian are exposed. According to the previous geological data, the study area strata are: Ordovician, Carboniferous, Permian, Quaternary (from old to new). Most of the coal seam in the field are in Benxi Formation, Taiyuan Formation and Shanxi Formation, among which the available coal seams are mainly distributed in Taiyuan Formation and Shanxi Formation. The thickness of 2# coal or 10# coal is relatively stable, and their reserves are considerable, they are the main mining coal seam in this region. The #10 coal in this study is in the lower section of Taiyuan K2 limestone, and is the main available coal seam in this area.

(2) **Geological structure**

The structure of the field belongs to the compound form of fault and fold. overall, the Wangjialing Mine is in a wing of a monoclinic structure. The monoclinic structure tends to north by west, and the sub-folds are developed on the basis of the monoclinic structure, and the sub-faults are developed, and which is mainly the positive faults. Stratigraphic dip angle is between 5° and 10°, but the angle in local areas can reach a

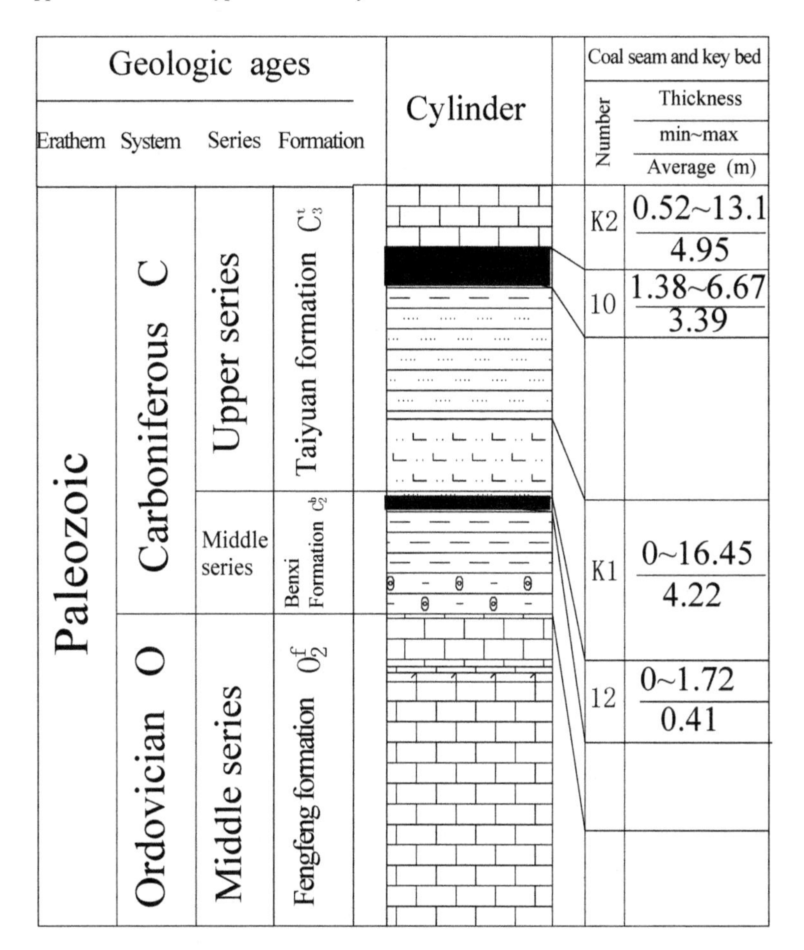

Fig. 5.15 Generalized columnar section of #10 seam's floor and aquifer

maximum of 37°. In the study area, the collapse column is rarely developed, and there is no intrusion of magma rock. As the structure of the study area is relatively less complex, so the formation is relatively complete. At present, there are two folds in the field, namely, S1 anticline and S2 syncline. 103 faults (positive faults) are found in the field, among which in the past ground spectrometry one fault was found, and the interpretation faults are 71, and the working face of the underground roadway expose the 31 faults. There is no magma rock in the field. Figure 5.15 shows the outline of the minefield structure.

5.2.2.2 Hydrogeological Background

Wangjialing Minefield is in the western runoff area of Yumenkouquan karst water system, and the Yumenkouquan karst water system constitutes a complete hydrogeological unit from recharge and runoff to discharge. In the study area, there are Ordovician limestone aquifer, Taiyuan limestone aquifer, Permian sandstone aquifer, Quaternary pore aquifer, aquifer and coal seam location relationship shown in Fig. 5.15, #10 Coal in this study is located at the bottom of K2 limestone in Taiyuan, and the interface of Ordovician is approximately 29.32 m from #10 coal floor. Because of the close distance between them, the Ordovician aquifer is the biggest threat to the safely mining of #10 coal.

5.2.3 Analysis and Determination of Pressure Area

The relationship between the standard water level of the Ordovician limestone and the elevation of #10 coal bottom is used to determines the Ordovician pressure area of #10 coal. The partition of the Ordovician pressure area is shown in the Fig. 5.16.

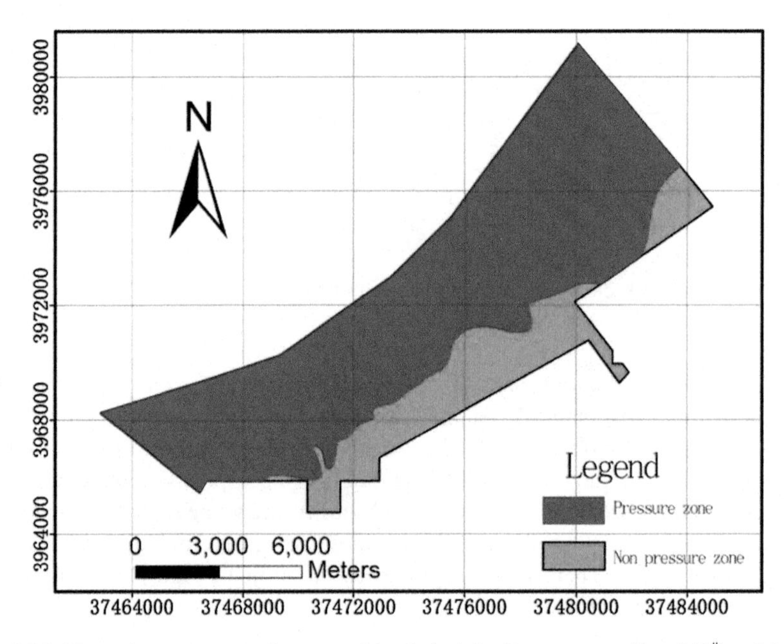

Fig. 5.16 The under-pressure zoning map of the Ordovician limestone aquifer of $10^{\#}$ coal floor

5.2.4 *Study on Main Controlling Factors of Ordovician Limestone Water Inrush #10 Coal Floor*

Based on the analysis of the hydrogeological conditions, the hydrogeological conditions in mine and the geological conditions of the Wangjialing Mine, this paper studies the main factors that influence floor water inrush of #10 coal from three aspects of the floor aquifer, the aquiclude and the tectonic situation. And ultimately determined that the eight factors: water abundance of aquiclude, the water pressure of the aquiclude, the equivalent thickness of the effective aquifers, the thickness of brittle rock under the mining pressure damage zone, the water resistance of the ancient weathering crust, the distribution of the fault and the fold axis, the distribution of end points and intersection of the fault and the fold axis, scale index of fault, are the main controlling factors influencing the floor water inrush of the 10# coal in Wangjialing Mine.

5.2.5 *Establishment of Thematic Map of Main Controlling Factor of the Floor Water Inrush*

After determining the main control factors, we first need to extract the data of the main control factors in the collected data, and use Surfer's interpolation function and GIS spatial analysis function to quantify the data, and then establish the main control factor thematic map Layer (Fig. 5.17). Because of the existence of the ancient weathering crust in the Ordovician limestone interface, this study believes that there is no Ordovician confined water raising belt; for the calculation of the depth of the mining pressure damage zone, because there is no measured data in research area, this paper takes empirical formula to calculate. In the study area there are many faults with uneven distribution. So, the faults are divided into 100 m $\times$ 100 m element meshes in the area where the fault is developed. In the area where the faults are not developed, they are divided into the 500 m $\times$ 500 m unit meshes.

5.2.6 *Vulnerable Evaluation of Water Inrush of #10 Coal Floor Based on ANN's Constant Weight Model*

5.2.6.1 Analytic Hierarchy Process (AHP) Model Design

(1) **Establish hierarchical analysis model**

Through the above analysis and the analysis of the hydrogeological conditions and geological conditions in the study area, the study object is divided into three levels (A, B and C) (Fig. 5.18).

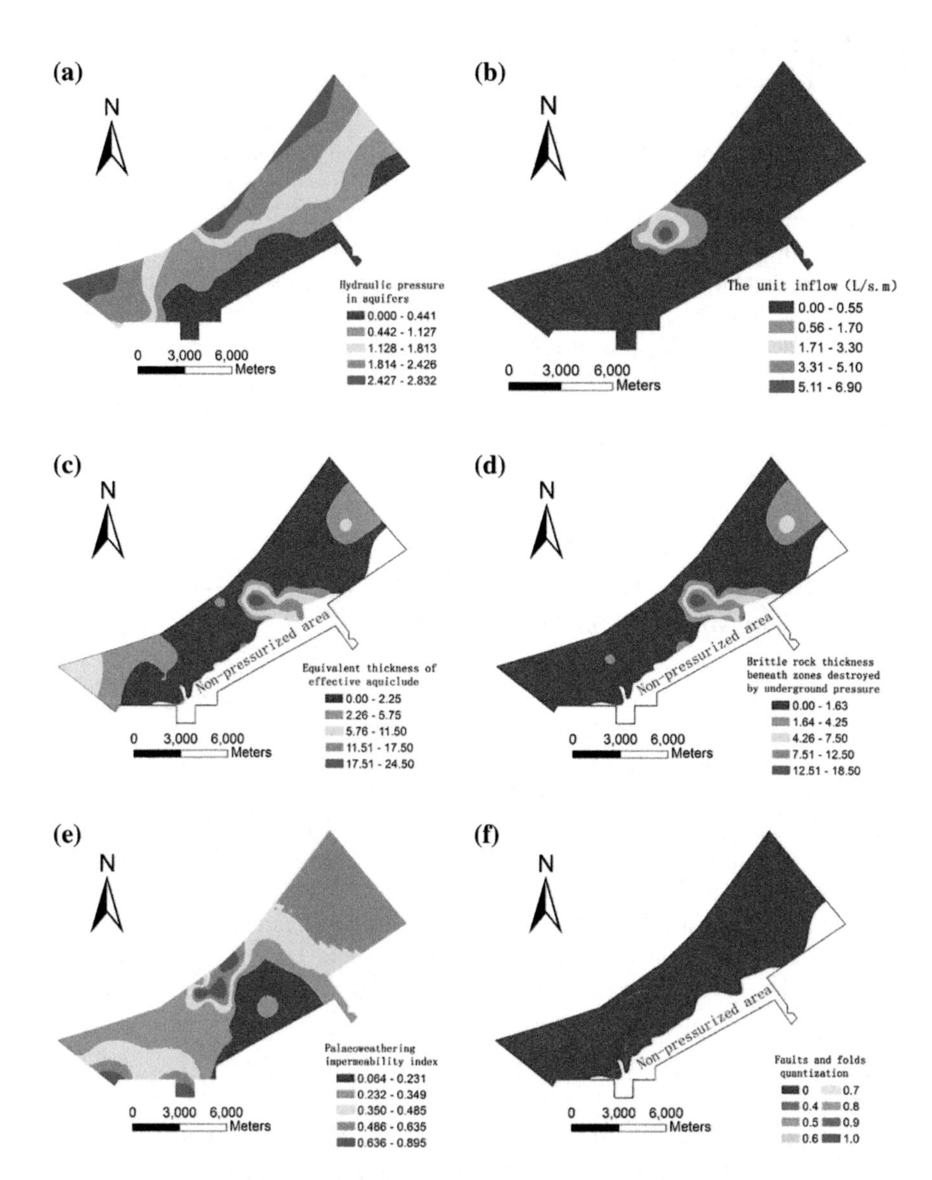

Fig. 5.17 Thematic layer graph of influencing factors that influence #10 coal seam floor water inrush [**a** the water pressure of Ordovician aquiclude; **b** the water abundance of the Ordovician aquiclude; **c** the equivalent thickness of the effective aquifers; **d** the thickness of the brittle rock below the floor mining pressure damage zone; **e** the impermeable performance of the ancient weathering crust; **f** the distribution of the fault and fold axis; **g** the distribution of the end points and intersections of fault; **h** scale index of fault]

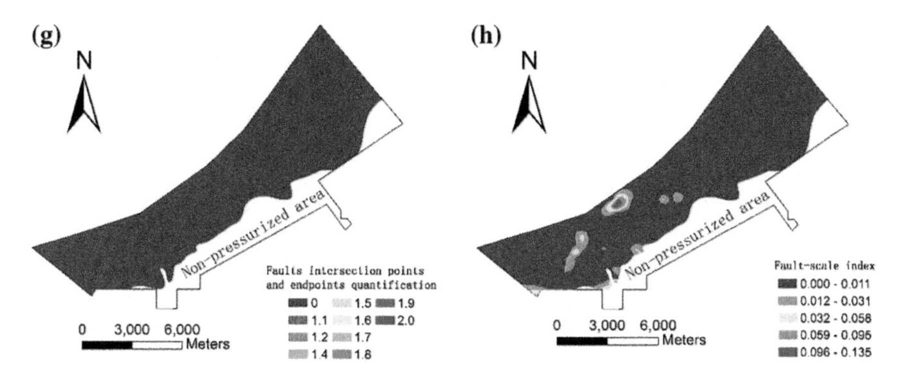

Fig. 5.17 (continued)

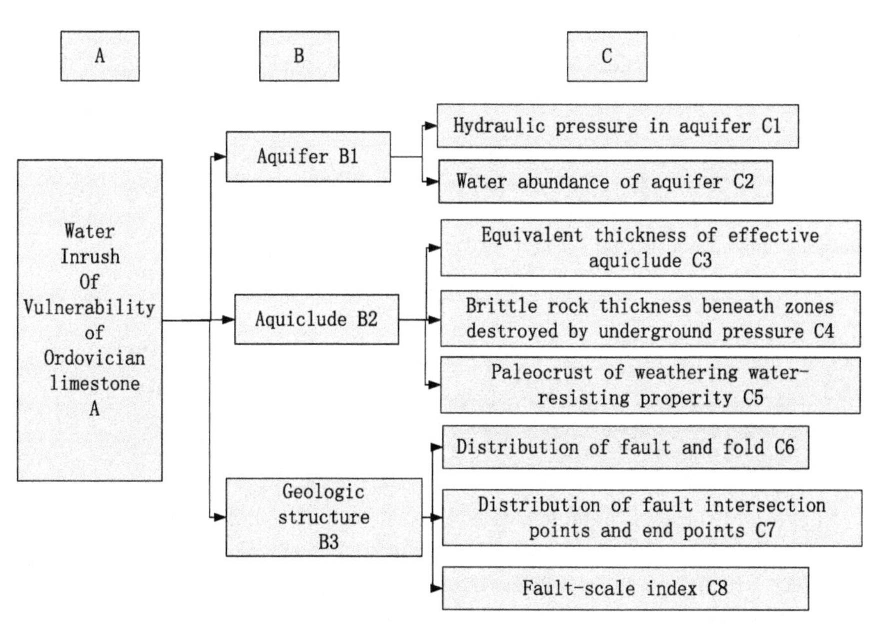

Fig. 5.18 Structure model of AHP of evaluation the vulnerability of water inrush of Ordovician limestone in mine floor

(2) Establish judgment matrix and consistency test

By consulting the recommendations and opinions of the relevant experts, relatively compare the “contribution” of the factors that influence the Ordovician limestone floor water inrush of the 10# coal in accordance with the 1–9 scale method. And on this basis to build judgement matrix (Tables 5.5, 5.6 and 5.7), and finally calculate the weight of the single order of each layer, that is, the W column in Table 5.8.

Table 5.5 Judgment matrix A − Bi (i = 1–3)

A	B_1	B_2	B_3	W(A/B)
B_1	1	1	2	0.4111
B_2	1	1	1	0.3278
B_3	1/2	1	1	0.2611

$\lambda_{max} = 3.0537, CI_1 = 0.02685, CR_1 = 0.0516 < 0.1$

Table 5.6 Judgment matrix B1 − Ci (i = 1–2)

B_1	C_1	C_2	W
C_1	1	3/2	0.6
C_2	2/3	1	0.4

$\lambda_{max} = 2, CI_{21} = 0, CR_{21} = 0 < 0.1$

Table 5.7 Judgment matrix B2 − Ci (i = 3–5)

B_2	C_3	C_4	C_5	W
C_3	1	3	4	0.6232
C_4	1/3	1	1/2	0.2395
C_5	1/4	1/2	1	0.1373

$\lambda_{max} = 3.0183, CI_{22} = 0.00915, CR_{22} = 0.0176 < 0.1$

Table 5.8 Judgment matrix B3 − Ci (i = 6–8)

B_3	C_6	C_7	C_8	$W(B_1/C_i)$
C_6	1	3	4	0.6327
C_7	1/3	1	1	0.1924
C_8	1/4	1	1	0.1749

$\lambda_{max} = 3.0092, CI_{23} = 0.0046, CR_{23} = 0.0089 < 0.1$

After establishing the judgment matrix, it is necessary to judge whether the established judgment matrix is consistent. According to the judgment principle, when the CR value of each group matrix is less than 0.1, then the established judgment matrix is reasonable. From the data listed in the table, it can be seen that the structure of the judgment matrix can pass the consistency test.

Based on the single-level sorting, the next step is to complete the A-level C-level hierarchy, which is shown in Table 5.9.

W (A/Ci) is the final decision data, and determine the weight of the eight main factors affecting the floor Ordovician limestone water inrush of the #10 coal seams (Table 5.10).

Table 5.9 Weights of different index to general goal

A/Ci	B1/0.32748	B2/0.4126	B3/0.25992	W(A/Ci)
C1	0.6			0.2467
C2	0.4			0.1644
C3		0.6232		0.2043
C4		0.2395		0.0785
C5		0.1373		0.045
C6			0.6327	0.1652
C7			0.1924	0.0502
C8			0.1749	0.0457

5.2.6.2 Vulnerability Evaluation of Water Inrush of #10 Coal Floor Based on AHP Constant Weight Model

(1) **Construction of thematic layer of data normalization and the main control of the normalization**

Due to the different dimensions of the main control factors, the dimensionality of the main control factors should be eliminated before the vulnerability evaluation, so that the different main control factors are comparable. The way to eliminate dimension is normalization, due to the different relativity of various main control factors and floor water inrush. Some main control factors are positive correlations, such as the water pressure of the aquiclude, the water abundance of the aquiclude, the distribution of fault and the folds axis, endpoints and intersecting point of fault, fault scale index; some main control factors are negative correlation factors, such as the equivalent thickness of the ancient weathering crust, the equivalent thickness of the effective aquifers, and the brittle rock thickness under the mining pressure damage zone. Therefore, in the normalization process, we should first determine the correlation between the main control factor and the floor water inrush. For the positive correlation factor, the maximum value method is used to normalize, and the negative correlation factor is normalized by the minimum method. The distribution of fault and fold axis and the distribution of the endpoints and intersection of fault and fold axis are normalized by eigenvalue method.

(2) **Establishment of vulnerability evaluation model and information fusion**

According to the constant weight determined by AHP, a vulnerability index method based on constant weight is established:

$$
\begin{aligned}
VI = \sum_{k=1}^{n} W_k \cdot f_k(x, y) &= 0.2467\mathrm{f}_1\,(\mathrm{x, y}) + 0.1644\mathrm{f}_2\,(\mathrm{x, y}) + 0.2043\mathrm{f}_3\,(\mathrm{x, y}) \\
&\quad + 0.0785\mathrm{f}_4\,(\mathrm{x, y}) + 0.045\mathrm{f}_5\,(\mathrm{x, y}) + 0.1652\mathrm{f}_6\,(\mathrm{x, y}) \\
&\quad + 0.0502\mathrm{f}_7\,(\mathrm{x, y}) + 0.0457\mathrm{f}_8\,(\mathrm{x, y})
\end{aligned}
$$

Table 5.10 Weights of different influencing factors that influence coal floor seam's water inrush

Influencing factors	Water pressure of aquifer (W1)	Water abundance of aquifer (W2)	Equivalent thickness of effective aquiclude (W3)	Thickness of brittle rock under mining pressure damage zone (W4)	Equivalent thickness of ancient weathering crust (W5)	Distribution of fault and fold (W6)	Distribution of endpoints and intersection of fault and fold (W7)	Scale index of fault (W8)
Weight (Wi)	0.2467	0.1644	0.2043	0.0785	0.045	0.1652	0.0502	0.0457

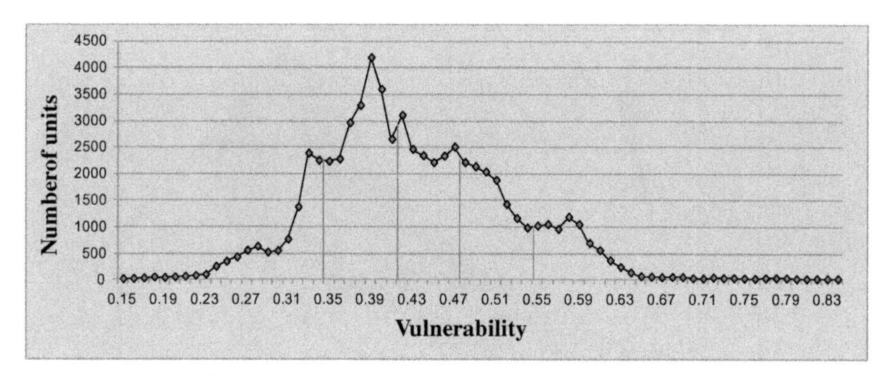

Fig. 5.19 Frequency histogram of vulnerable index of mine floor

where: *VI* is vulnerability index;
W_k is the weight of main control factors;
$f_k(x,y)$ is the single factor influence value function;
x,y is geographical coordinates;
n is the number of influence factor.

Then, the non-dimensional layers processed by main control factors are processed and analyzed by GIS, and new topological relations are formed according to different attributes, and the number of units of vulnerability index is further determined.

(3) **Partition of coal seam floor water inrush vulnerability evaluation**

Natural Breaks is used to study the vulnerability index rate statistical graph in each unit of segment, and the thresholds of vulnerability evaluation of 10# coal seam Ordovician limestone water inrush are as follows: 0.34, 0.41, 0.47, 0.54 (Fig. 5.19). The size of the vulnerability index can directly reflect the size of the possibility of water inrush. The study area is divided into five regions according to the partition threshold:

$VI \geq 0.54$ Coal seam floor water inrush vulnerable zone
$0.47 < VI \leq 0.54$ Coal seam floor water inrush more vulnerable zone
$0.41 < VI \leq 0.47$ Coal seam floor water inrush transition zone
$0.34 < VI \leq 0.41$ Coal seam floor water bursts less safe zone
$VI < 0.34$ Coal seam floor water inrush relative safe zone.

After determining the threshold of the partition, it is possible to obtain the partition map of the vulnerability evaluation of the Ordovician limestone water inrush in the #10 coal floor (Fig. 5.20). As the current 10# coal mine has not yet mined, so it randomly selected 10 points to have fitting verification on the evaluation results, the final selection of 10 fitting points coincides with the actual, the fitting rate reaches 100%. Therefore, the existed vulnerability evaluation partition has a higher credibility, and the evaluation results are ideal.

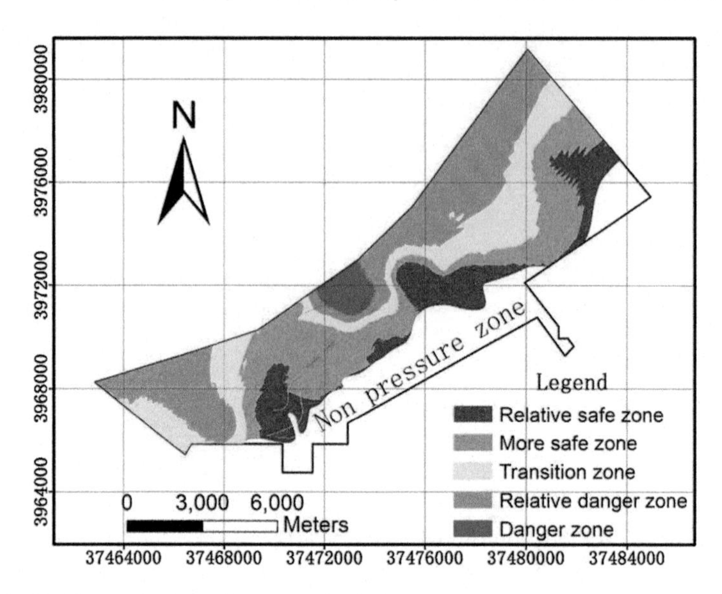

Fig. 5.20 Division evaluation of the vulnerable index method

Fig. 5.21 State variable weight vector of different main controlling factors

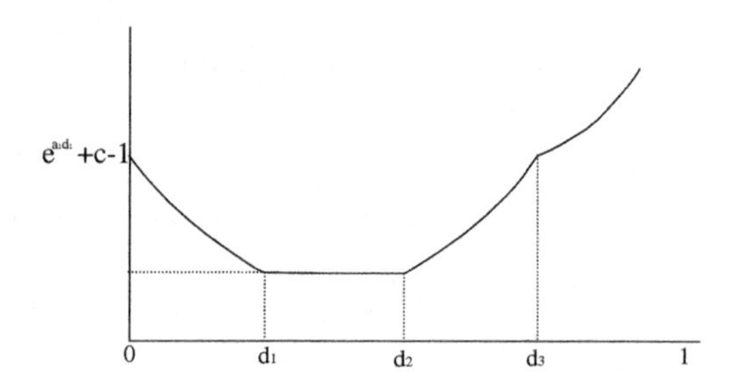

It can be seen from Fig. 5.21 that the risk of water inrush in the Ordovician limestone aquifer can be divided into five different grades. In general, the risk of Ordovician limestone water inrush in the western part and middle-northern part of the #10 coal in study area is relatively large, and the risk of Ordovician limestone water inrush in the southern and east-southern part of the study area is relatively small.

5.2.7 Vulnerable Evaluation of Water Inrush of 10# Coal Floor Based on Partition Variable Weight Model

(1) **Partition state variable weight vector construction**

According to the variable weight of evaluation characteristics of coal seam floor water inrush, the variable weight model can reflect the influence of the sudden change of the main control factor value on the evaluation result more accurately. According to the positive and negative correlation between the index value and the floor water inrush, the "penalty" and "incentive" measures were taken to further highlight or weaken the impact on the floor water inrush. In this study, the state variable vector mathematical model is determined as follows. The state vector curve is shown in Fig. 5.21.

Since the contribution of different indicators to the floor water inrush is different, the state value is divided into four intervals, namely, the strong excitation interval [d3, 1], the initial excitation interval [d2, d3), the non-penalty non-excitation interval [d1, d2), punishment interval [0, d1).

$$S_j(x) = \begin{cases} e^{a_1(d_1-x)} + c - 1, & x \in [0, d_1) \\ c, & x \in [d_1, d_2) \\ e^{a_2(x-d_2)} + c - 1, & x \in [d_2, d_3) \\ e^{a_3(x-d_3)} + e^{a_2(d_3-d_2)} + c - 2, & x \in [d_3, 1) \end{cases}$$

where a_1, a_2, a_3, c are the parameters to be determined and d1, d2, d3 are the variable weight range thresholds of the main control factors.

(2) **Determination of main control factors variable weight range and the adjust weight parameters**

In this evaluation, the K-means clustering method in dynamic clustering is used to analyze the factor index value. The clustering method selects the iterative classification according to the K-means algorithm. According to the constructed variable weight model, the determined classification category is 4 categories. And, then the water pressure, the thickness of the aquifers and other eight factors affecting the floor water inrush were divided respectively, and get the index value classification critical value of main control factors, after obtaining the average value the variable weight range threshold can be achieved. For the two main control factors of the fault distribution and the distribution of endpoints and intersection of the fault and the fold, their index value is fixed and less. In this evaluation, the various weight range threshold of the two factors is determined by the previous application experience. And finally get the corresponding main control variable range (Table 5.11).

In this study, after repeated analysis and debugging, when achieve the ideal variable effect, $a_1 = 0.6923$, $a_2 = 0.8234$, $a_3 = 0.4681$, $c = 0.5216$.

Table 5.11 Variable weight intervals of main controlling factors

Main control factor	Property of interval range			
	Punishment interval	Non-penalty non-excitation interval	Initial excitation interval	Strong excitation interval
Water pressure of aquifer	$0.208 > x \geq 0$	$0.381 > x \geq 0.208$	$0.623 > x \geq 0.381$	$1 \geq x \geq 0.623$
Coal floor equivalent thickness of effective aquiclude	$0.469 > x \geq 0$	$0.755 > x \geq 0.469$	$0.939 > x \geq 0.755$	$1 \geq x \geq 0.939$
Brittle rock thickness under mining pressure damage zone	$0.405 > x \geq 0$	$0.723 > x \geq 0.405$	$0.919 > x \geq 0.723$	$1 \geq x \geq 0.919$
Water abundance of aquiclude	$0.06 > x \geq 0$	$0.203 > x \geq 0.06$	$0.493 > x \geq 0.203$	$1 \geq x \geq 0.494$
Distribution of fault		$0.500 > x \geq 0$	$0.800 > x \geq 0.500$	$1 \geq x \geq 0.800$
Distribution of endpoints and intersection points of fault and fold		$0.350 > x \geq 0$	$0.500 > x \geq 0.350$	$1 \geq x \geq 0.500$
Scale index of fault	$0.093 > x \geq 0$	$0.278 > x \geq 0.093$	$0.593 > x \geq 0.278$	$1 \geq x \geq 0.593$
Water resistance of ancient weathering crust on the Ordovician top	$0.416 > x \geq 0$	$0.640 > x \geq 0.416$	$0.778 > x \geq 0.640$	$1 \geq x \geq 0.778$

(3) **Calculation of the variable weight of the main control factors**

After establishing the partitioned variable weight model of the main control factor of 10# coal seam floor water inrush, and then use the partitioned variable weight model to calculate the weight of the main control factor that change with the change of index value (Table 5.12).

The variable weights of the main control factors are determined by using the partitioned variable weight model. The mathematical expression is as follows:

$$W(X)=\frac{W_0 \cdot S(X)}{\sum_{j=1}^{8} w_j^{(0)} S_j(X)}=\left(\frac{w_1^{(0)} S_1(X)}{\sum_{j=1}^{8} w_j^{(0)} S_j(X)}, \frac{w_2^0 S_2(X)}{\sum_{j=1}^{8} w_j^0 S_j(X)}, \ldots, \frac{w_8^0 S_8(X)}{\sum_{j=1}^{8} w_j^0 S_j(X)}\right)$$

where: $S(X)$ are state variable vectors;
$W_0 = \left(w_1^{(0)}, w_2^{(0)}, \ldots, w_8^{(0)}\right)$ are weights;
$W(X)$ are partition variable vectors.

Taking the water pressure of the aquiclude as an example, we get the variable weight analysis chart of the water pressure in the aquiclude. It can be seen that the change of the water pressure weight of the aquifer is basically the same as the established state vector trend.

(4) **Establishment of vulnerability index method of partitioned variable weight model**

Based on the weight of variable weight of main control factors determined by partitioned variable weight model, under the premise of fully analysis of basic hydrogeological conditions, further construct the basic mathematical model of vulnerability evaluation of 10# coal floor Ordovician limestone water inrush of Wangjialing coal mine. It is shown in the formula:

$$\begin{aligned} VI = \sum_{i=1}^{8} w_i(z) \cdot f_i(x, y) = \sum_{i=1}^{8} \frac{w_i^{(0)} S_i(X)}{\sum_{j=1}^{8} w_j^{(0)} S_j(X)} f_i(x, y) &= \frac{w_1^{(0)} S_1(X)}{\sum_{j=1}^{8} w_j^{(0)} S_j(X)} f_1(x, y) \\ &+ \frac{w_2^{(0)} S_2(X)}{\sum_{j=1}^{8} w_j^{(0)} S_j(X)} f_2(x, y) + \frac{w_3^{(0)} S_3(X)}{\sum_{j=1}^{8} w_j^{(0)} S_j(X)} f_3(x, y) \\ &+ \frac{w_4^{(0)} S_4(X)}{\sum_{j=1}^{8} w_j^{(0)} S_j(X)} f_4(x, y) + \frac{w_5^{(0)} S_5(X)}{\sum_{j=1}^{8} w_j^{(0)} S_j(X)} f_5(x, y) \\ &+ \frac{w_6^{(0)} S_6(X)}{\sum_{j=1}^{8} w_j^{(0)} S_j(X)} f_6(x, y) + \frac{w_7^{(0)} S_7(X)}{\sum_{j=1}^{8} w_j^{(0)} S_j(X)} f_7(x, y) \\ &+ \frac{w_8^{(0)} S_8(X)}{\sum_{j=1}^{8} w_j^{(0)} S_j(X)} f_8(x, y) \end{aligned}$$

Table 5.12 Variable weights of main controlling factors

Water abundance	Intersections of faults and folders	Friable rock	Distribution of faults and folders	Water pressure	Fault scale index	Equivalent thickness of aquiclude	Weathering crust
0.196983	0.037358	0.082316	0.122939	0.224768	0.035829	0.263848	0.035959368
0.194669	0.036919	0.081349	0.121495	0.233873	0.035408	0.26075	0.035537049
0.194528	0.037524	0.082681	0.123484	0.225766	0.035988	0.265019	0.035011
0.194312	0.037482	0.082589	0.123348	0.225516	0.035948	0.264726	0.036078947
0.192084	0.037689	0.083044	0.124027	0.226758	0.036146	0.266183	0.034070427
0.191874	0.037647	0.082953	0.123891	0.22651	0.036106	0.265892	0.035126362
0.154404	0.042183	0.090837	0.138819	0.26886	0.043298	0.222241	0.039358645
0.154212	0.042131	0.090724	0.138646	0.268525	0.043244	0.221964	0.040553635
0.154017	0.042078	0.090609	0.138471	0.268186	0.04319	0.221684	0.041765682
0.15457	0.042229	0.091088	0.138967	0.269148	0.043345	0.222479	0.038174641
0.148385	0.042358	0.091212	0.139393	0.269971	0.043477	0.22316	0.042043725
0.148219	0.04231	0.091264	0.139236	0.269669	0.043428	0.22291	0.042963299
0.148954	0.04252	0.09182	0.139927	0.271006	0.043644	0.224015	0.038115602
0.148906	0.042506	0.09179	0.139881	0.270918	0.04363	0.223943	0.03842574
0.148204	0.042306	0.091357	0.139222	0.269641	0.043424	0.222887	0.042958898
0.148107	0.042278	0.091298	0.139131	0.269466	0.043396	0.222742	0.043581672
0.148931	0.042513	0.09196	0.139905	0.270964	0.043637	0.22398	0.038109698
0.148931	0.042513	0.09196	0.139905	0.270964	0.043637	0.22398	0.038109698
0.148085	0.042272	0.091438	0.13911	0.269424	0.043389	0.222707	0.043574959
0.148885	0.0425	0.092242	0.139862	0.27088	0.043623	0.223911	0.038097886
0.149334	0.042629	0.091796	0.140284	0.271697	0.043755	0.218193	0.04231255

(continued)

Table 5.12 (continued)

Water abundance	Intersections of faults and folders	Friable rock	Distribution of faults and folders	Water pressure	Fault scale index	Equivalent thickness of aquiclude	Weathering crust
0.142601	0.042571	0.092396	0.140096	0.271333	0.043696	0.229145	0.038161629
0.142601	0.042571	0.092396	0.140096	0.271333	0.043696	0.229145	0.038161629
0.142542	0.042554	0.092773	0.140037	0.27122	0.043678	0.22905	0.038145786
0.142542	0.042554	0.092773	0.140037	0.27122	0.043678	0.22905	0.038145786
0.142522	0.042548	0.092035	0.140018	0.271183	0.043672	0.224162	0.043859477
0.142381	0.042506	0.091944	0.139879	0.270914	0.043629	0.223939	0.044807614
0.142337	0.042493	0.092225	0.139836	0.27083	0.043615	0.22387	0.044793729
0.139469	0.042587	0.092846	0.140147	0.271433	0.043713	0.23163	0.038175705
0.139469	0.042587	0.092846	0.140147	0.271433	0.043713	0.23163	0.038175705
0.125399	0.038291	0.083106	0.12601	0.325498	0.039303	0.206106	0.05628773
0.125353	0.038277	0.083449	0.125962	0.325376	0.039288	0.206029	0.056266712
0.138927	0.042422	0.092071	0.139603	0.270378	0.043543	0.228339	0.044718908
0.138741	0.042365	0.091948	0.139416	0.270016	0.043484	0.228033	0.045996285
0.139804	0.04269	0.093069	0.140484	0.272086	0.043818	0.229781	0.038267565
0.139804	0.04269	0.093069	0.140484	0.272086	0.043818	0.229781	0.038267565
0.138683	0.042347	0.092323	0.139358	0.269905	0.043466	0.227939	0.045977282
0.139603	0.042628	0.092519	0.140282	0.271694	0.043754	0.224584	0.044936525
0.139186	0.042501	0.093074	0.139864	0.270884	0.043624	0.232768	0.038098492
0.139469	0.042587	0.092846	0.140147	0.271433	0.043713	0.23163	0.038175705
0.13941	0.042569	0.093224	0.140089	0.27132	0.043694	0.231534	0.038159801
0.13941	0.042569	0.093224	0.140089	0.27132	0.043694	0.231534	0.038159801

(continued)

Table 5.12 (continued)

Water abundance	Intersections of faults and folders	Friable rock	Distribution of faults and folders	Water pressure	Fault scale index	Equivalent thickness of aquiclude	Weathering crust
0.138109	0.042172	0.092353	0.138781	0.268787	0.043286	0.229372	0.047140262
0.138683	0.042347	0.092323	0.139358	0.269905	0.043466	0.227939	0.045977282
0.138626	0.04233	0.092699	0.139301	0.269793	0.043448	0.227845	0.045958236
0.138438	0.042272	0.092573	0.139112	0.269427	0.043389	0.227536	0.047252584
0.138109	0.042172	0.092353	0.138781	0.268787	0.043286	0.229372	0.047140262
0.139186	0.042501	0.093074	0.139864	0.270884	0.043624	0.232768	0.038098492
0.138919	0.042419	0.093207	0.139595	0.270363	0.04354	0.233932	0.038025206
0.13883	0.042392	0.093148	0.139506	0.27019	0.043512	0.233782	0.038639722
0.139186	0.042501	0.093074	0.139864	0.270884	0.043624	0.232768	0.038098492
0.139143	0.042488	0.093358	0.13982	0.270799	0.04361	0.232695	0.038086572
0.138066	0.042159	0.092635	0.138738	0.268703	0.043273	0.229301	0.047125627
0.138044	0.042152	0.092776	0.138716	0.268661	0.043266	0.229265	0.047118299
0.13883	0.042392	0.093148	0.139506	0.27019	0.043512	0.233782	0.038639722
0.137825	0.042085	0.092629	0.138496	0.268234	0.043197	0.230491	0.047043338
0.137811	0.042081	0.092722	0.138481	0.268206	0.043193	0.230467	0.047038476
0.137803	0.042079	0.092769	0.138474	0.268192	0.043191	0.230455	0.047036043
0.138044	0.042152	0.092776	0.138716	0.268661	0.043266	0.229265	0.047118299
0.13883	0.042392	0.093148	0.139506	0.27019	0.043512	0.233782	0.038639722
0.138808	0.042386	0.09329	0.139484	0.270148	0.043506	0.233746	0.03863368
0.137803	0.042079	0.092769	0.138474	0.268192	0.043191	0.230455	0.047036043

(continued)

Table 5.12 (continued)

Water abundance	Intersections of faults and folders	Friable rock	Distribution of faults and folders	Water pressure	Fault scale index	Equivalent thickness of aquiclude	Weathering crust
0.138817	0.042388	0.093295	0.139493	0.270165	0.043445	0.233761	0.038636118
0.138835	0.042394	0.093307	0.13951	0.270199	0.043324	0.23379	0.038641004
0.138861	0.042402	0.093325	0.139537	0.27025	0.043143	0.233834	0.038648312
0.138846	0.042397	0.093419	0.139522	0.270222	0.043139	0.23381	0.038644288
0.138881	0.042408	0.093443	0.139557	0.27029	0.042898	0.233869	0.038654002
0.137584	0.042012	0.092621	0.138253	0.267764	0.043122	0.231683	0.046960986
0.137395	0.041954	0.092494	0.138064	0.267398	0.043063	0.231366	0.048265822
0.137803	0.042079	0.092769	0.138474	0.268192	0.043191	0.230455	0.047036043
0.138881	0.042408	0.093443	0.139557	0.27029	0.042898	0.233869	0.038654002
0.138748	0.042367	0.093353	0.139423	0.27003	0.042607	0.234856	0.038616831
0.138741	0.042365	0.0934	0.139416	0.270016	0.042604	0.234844	0.03861482
0.138916	0.042418	0.093466	0.139592	0.270358	0.042658	0.233927	0.038663691
0.138782	0.042378	0.093376	0.139458	0.270097	0.042368	0.234914	0.03862647
0.138775	0.042375	0.093423	0.13945	0.270083	0.042366	0.234902	0.038624458
0.138898	0.042413	0.093506	0.139574	0.270323	0.042154	0.235111	0.038019641
0.13881	0.042386	0.093447	0.139485	0.270151	0.042127	0.234961	0.03863407
0.138924	0.042421	0.093524	0.1396	0.270374	0.041976	0.235155	0.038026712
0.138941	0.042426	0.093535	0.139618	0.270407	0.041857	0.235184	0.038031432
0.138967	0.042434	0.093553	0.139644	0.270457	0.041679	0.235227	0.038038494
0.139002	0.042445	0.093576	0.139678	0.270524	0.041443	0.235285	0.038047882

(continued)

Table 5.12 (continued)

Water abundance	Intersections of faults and folders	Friable rock	Distribution of faults and folders	Water pressure	Fault scale index	Equivalent thickness of aquiclude	Weathering crust
0.139036	0.042455	0.093599	0.139712	0.270591	0.041207	0.235343	0.038057245
0.13907	0.042465	0.093622	0.139747	0.270657	0.040971	0.235401	0.038066583
0.139019	0.04245	0.093588	0.139696	0.270558	0.040712	0.235924	0.038052714
0.139104	0.042476	0.093645	0.139781	0.270723	0.040737	0.235459	0.038075895
0.13907	0.042465	0.093622	0.139747	0.270657	0.040362	0.23601	0.03806661
0.137395	0.041954	0.092494	0.138064	0.267398	0.043063	0.231366	0.048265822
0.138917	0.042419	0.093519	0.139593	0.27036	0.041417	0.235751	0.038024735
0.138951	0.042429	0.093542	0.139627	0.270426	0.041182	0.235809	0.038034087
0.138985	0.04244	0.093565	0.139662	0.270492	0.040947	0.235866	0.038043413
0.139019	0.04245	0.093588	0.139696	0.270558	0.040712	0.235924	0.038053
0.13907	0.042465	0.093622	0.139747	0.270657	0.040362	0.23601	0.038067
0.13907	0.042465	0.093622	0.139747	0.270657	0.040362	0.23601	0.038067
0.138122	0.04186	0.092288	0.137756	0.266801	0.042967	0.232048	0.048158
0.138287	0.041911	0.092399	0.137921	0.267121	0.043018	0.231127	0.048216
0.139675	0.042331	0.093326	0.139305	0.269801	0.042321	0.234657	0.038584
0.139709	0.042342	0.093349	0.139339	0.269868	0.042083	0.234715	0.038594
0.139735	0.04235	0.093366	0.139365	0.269918	0.041905	0.234759	0.038601
0.139753	0.042355	0.093378	0.139383	0.269952	0.041787	0.234788	0.038606
0.139842	0.042382	0.093438	0.139472	0.270124	0.041813	0.234938	0.037992
0.139868	0.04239	0.093455	0.139497	0.270174	0.041636	0.234981	0.037999

(continued)

Table 5.12 (continued)

Water abundance	Intersections of faults and folders	Friable rock	Distribution of faults and folders	Water pressure	Fault scale index	Equivalent thickness of aquiclude	Weathering crust
0.139817	0.042374	0.093421	0.139447	0.270077	0.041374	0.235504	0.037985
0.139902	0.0424	0.093478	0.139532	0.270241	0.041399	0.235039	0.038008
0.139264	0.041805	0.092166	0.137573	0.266448	0.04291	0.23174	0.048094
0.14049	0.042173	0.092977	0.138784	0.268793	0.043099	0.23378	0.039905
0.140516	0.042181	0.092994	0.13881	0.268843	0.042919	0.233824	0.039913
0.140722	0.042243	0.093131	0.139014	0.269238	0.042982	0.234167	0.038504
0.140758	0.042253	0.093154	0.139049	0.269305	0.042742	0.234226	0.038513
0.140793	0.042264	0.093178	0.139083	0.269373	0.042503	0.234284	0.038523
0.140828	0.042274	0.093201	0.139118	0.26944	0.042265	0.234342	0.038532
0.140325	0.041724	0.091988	0.137308	0.265934	0.042827	0.231892	0.048002
0.141616	0.042108	0.092835	0.138572	0.268382	0.043221	0.233422	0.039844
0.140409	0.041749	0.092043	0.13739	0.266093	0.042853	0.231432	0.04803
0.141625	0.042111	0.092841	0.13858	0.268398	0.043161	0.233437	0.039846
0.141472	0.041669	0.091865	0.137125	0.265579	0.04277	0.231582	0.047938

Note Only portion of the data provided here

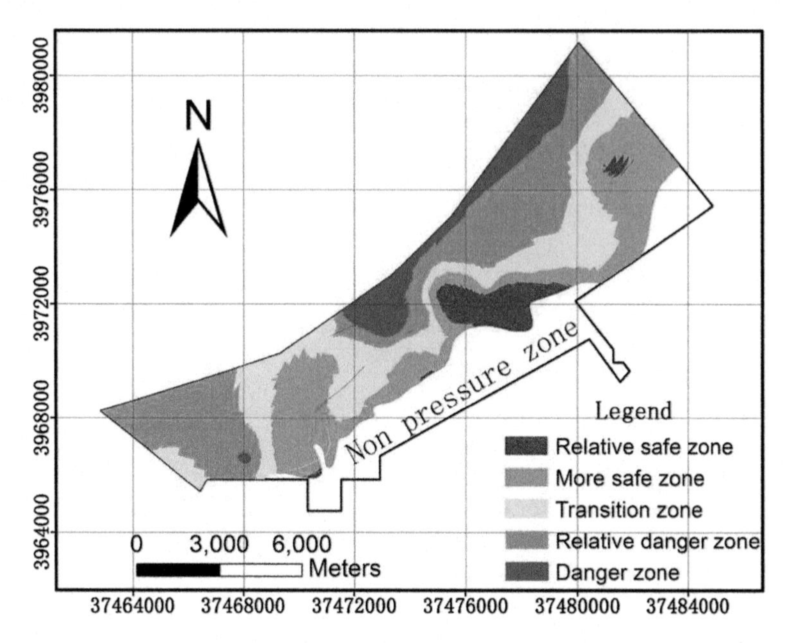

Fig. 5.22 Water inrush vulnerability assessment zoning map of 10# coal seam floor to Ordovician limestone aquifer

where, VI—vulnerability index;
Wi—variable weight vector of influencing factor;
W(0)—any constant weight vector;
fi(x,y)—the i th single factor influencing index function;
(x,y)—Geographical coordinates
$S(X) = \{S_1(X), S_2(X), \ldots S_8(X)\}$—8—dimensional partition state variable weight vector.

(5) **Partition of vulnerability evaluation of coal floor water inrush based on partitioned variable weight model**

The vulnerability index (VI) of each stacking unit in the study area is calculated by the above formula. The greater the value, the more possible having the occurrence of water inrush. Then we use the Natural Breaks (Jenks) commonly used on the grading map to classify the vulnerability index values, and then determine the threshold of the partition 0.34, 0.42, 0.50, 0.59. Then the evaluation partition map of vulnerability index method of coal floor Ordovician limestone water inrush is obtained (Fig. 5.22).

$VI > 0.59$ Coal seam floor water inrush vulnerable zone
$0.50 < VI \leq 0.59$ Coal seam floor water inrush more vulnerable zone
$0.42 < VI \leq 0.50$ Coal seam floor water inrush transition zone
$0.34 < VI \leq 0.42$ Coal seam floor water bursts safer zone
$VI \leq 0.34$ Coal seam floor water inrush relative safe zone.

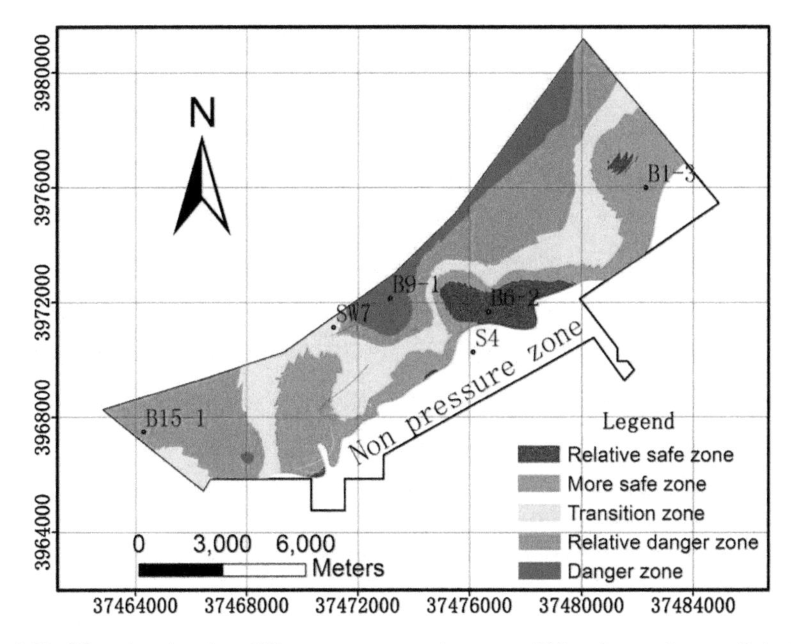

Fig. 5.23 Water inrush vulnerability assessment zoning map of 9# coal seam floor to Ordovician limestone aquifer

(6) **Identification and testing of model**

10# coal seam is not yet mined, there is no actual water point data to fit, here select B9-1, B15-1, SW7, B1-3, B6-2, S4 six drilling as a fitting point. The evaluation partitioned fitting graph is shown in Fig. 5.23. Through the verification and analysis, the selected fitting points are in the same positions as the evaluation result, so the evaluation result of this model is ideal.

5.2.8 Comparative Analysis of Variable Weight Model and Constant Weight Model and Water Inrush Coefficient Method

5.2.8.1 Comparative Analysis of the Evaluation Results of Variable Weight Model and Constant Weight Model

Compare the partitioned resulting map (Fig. 5.24) of the vulnerability evaluation on $10^{\#}$ coal floor water inrush of variable weight model and traditional constant weight model. When the two models were combined to conduct comparative analysis, the results of the two evaluations are matched in most areas. The possibility of water inrush from the southeast to north and northwest of the study area is increasing.

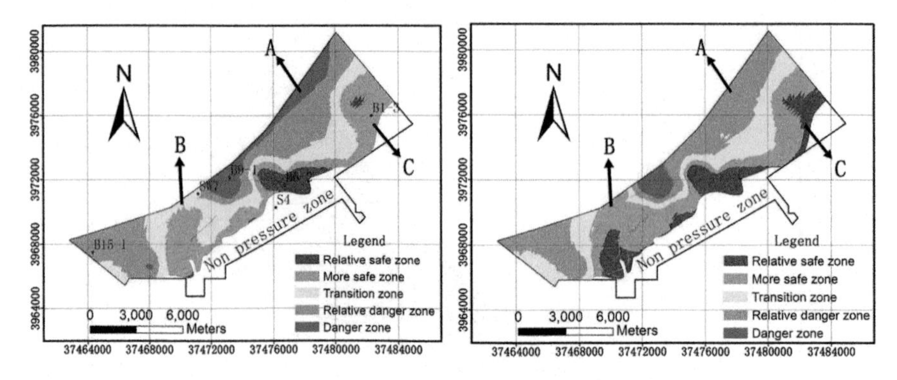

Fig. 5.24 Vulnerability assessment comparison graph based on variable weight model (left map) and constant weight model (right map)

However, it is also evident that the evaluation results of local areas are different. Here we select the region (A, B, C area) with the representative local differences for comparison.

Combine both to analyze, it is not difficult to find that compared with traditional constant weight model, the variable weight model can better reflect the influence of the change of the main control factor value on the evaluation result, and the evaluation result is more in line with the actual situation of the threat of mine floor water damage and more accurate. As in the A area above, the "incentive" indicator value contributes to the floor water burst. The water pressure in this area is relatively large and the water resistance is small, and the change of these two factors plays a role in promoting the water inrush. Therefore, the variable weight model gives appropriate "incentive" to the water pressure and the weight of thickness of the aquifer, it shows in that the orange more vulnerable area of constant weight model turns into the red vulnerable area of variable weight model. As in the B region above, although the equivalent thickness of the ancient weathering crust is larger, the distribution quantitation value of the fault and the fold axis is lower, but the unit water inflow is larger, and the change of the factor also promote water inrush. It shows that the light green more safety area in the evaluation result of the constant weight model becomes the yellow transition area in the evaluation results of the variable weight model. In the same way, compared with the C zone in the figure, the equivalent thickness of the effective aquifer in the area and the thickness of the brittle rock in the damaged fault zone are getting small, it changed from the relative safety zone in the constant weight model to the more safety area in variable weight model. Therefore, the variable weight model effectively avoids the control index being neutralized by other indicators. At the same time, the variable weight model constructed in this study also fully considered the factors that have hinder effect on the floor water inrush. To the factor that obviously impede floor water inrush, we also made a relative increase on their weight.

5.2.8.2 Comparison and Analysis of the Evaluation Results of Variable Weight Model and Water Inrush Method

In addition, the traditional water inrush coefficient method is used to evaluate the risk of water inrush in #10 coal seam floor Ordovician limestone water inrush. Due to the tectonic development of the study area, the critical water inrush coefficient is 0.06 MPa/m, and the water inrush risk of the coal seam floor is divided by the water inrush coefficient with a threshold of 0.06. The comparison between the variable weight model and the water inrush coefficient method is shown in Fig. 5.25.

By comparing the evaluation results of the partition variable weight model of #10 coal floor water inrush in Wangjialing mine field and the results of the traditional water inrush coefficient:

[1] The coal seam floor Ordovician limestone aquiclude water inrush vulnerability index method based on the partitioned variable weight model fully considered the eight main control factors of the aquifer, the aquiclude and the geological structure and analyzed and evaluated coal seam floor aquiclude. Therefore, the vulnerability evaluation results include richer information, and more comprehensive considered factors. But the water inrush coefficient method only considered the two factors of the Ordovician limestone water pressure and the thickness of the aquifer, thus ignores the influence of other factors, which is obviously not correspond to the actual situation. So, it is difficult to accurately evaluate the risk of water inrush from a comprehensive point of view.

[2] The evaluation accuracy of the water inrush coefficient method is poor, the evaluation result is only divided into the safety zone and the danger zone, and the vulnerability index method has five levels, which is more instructive to the actual.

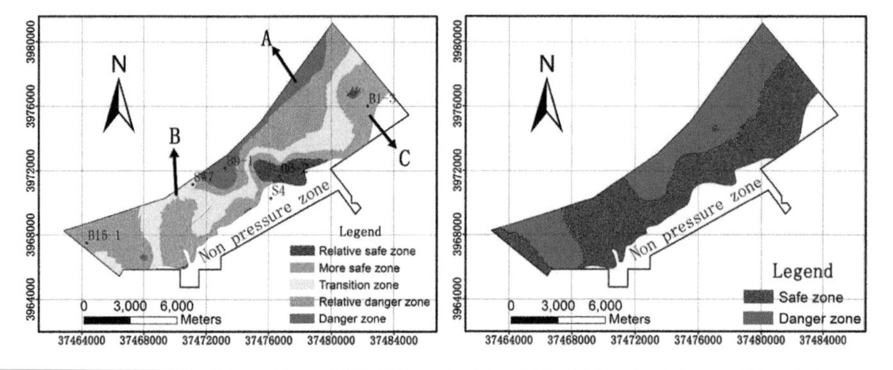

Fig. 5.25 Vulnerability assessment comparison graph based on variable weight model and water bursting coefficient method

[3] From the comparative analysis of hydrogeological evaluation formula: the water inrush coefficient method is only that conclusion that considers the two aspects of the thickness of the floor of the coal seam and the head pressure of the bottom of the aquifer, between which is only a ratio having no "weight" concept. But the vulnerability evaluation of coal seam floor of the variable weight model is from a comprehensive consideration of eight main control factors from aquiclude, geological structures and aquifers, and the variable weight model is used to reflect the influence of sudden change of factor index on the risk of water inrush. By comparison, the variable rights involve more factors, and its evaluation results are more in line with the actual situation.

5.3 Summary

In this chapter, it mainly uses the vulnerability index method based on the theory of partition variable weight to evaluate the vulnerability of #9 coal floor Ordovician limestone water inrush in "monoclinic order type" Xiandewang Coal Mine, and the vulnerability of $10^{\#}$ coal floor Ordovician limestone water inrush in "monoclinic inversion type" Wangjialing Coal Mine. The evaluation results are compared and analyzed with the evaluation results of traditional water inrush coefficient method and vulnerability index method based on the constant weight. The results show that the vulnerability index method based on constant weight theory can not only consider a variety of main control factors and their corresponding different weights, but also can quantitatively determine the corresponding weight of the same factor in different state values, the evaluation results are more realistic.

Chapter 6
Conclusion and Outlook

6.1 Conclusions

Based on the regularity and characteristics of coal formation, water filling source and water filling mode, and according to the geological structure of coal-bearing strata and superposition relation of water filling source, this paper divides karst water system and the geological structure model of coal-bearing stratum in the main coal mine area in north China into five types. The main factors that influence the water damage of the floor are analyzed, and the methods of collecting, quantifying and dimensioning the main control factors are systematically summarized, and the main control system of floor water inrush is established. Information fusion technology is used to establish a JDL-type water inrush evaluation information fusion model based on GIS. On this basis, the theory and method of floor water inrush evaluation based on partitioned variable weight theory are studied by using variable weight theory. And the vulnerability index method based on the theory of partitioned variable weight is applied to evaluate the risk of water inrush on coal seam floor in Xiandewang coal mine of monoclinic order type and Wangjialing coal mine, which is a monoclinic inversion type. The main research results are as follows:

[1] A classification method for spatial superposition of coal-bearing stratum and underlying karst water system is proposed. Based on the coal formation, water filling source and water filling method, the sedimentary regularity of the main coal-bearing Carboniferous coal seam in the north China coalfields is analyzed. The most important water damage in the north China coalfields and characteristics of karst water filling in Cambrian Ordovician Limestone are studied. And it summarizes the tectonic water control functions of regional structure, mining structure and field structure on the north China type coalfields. According to the superposition of the geological structure of the coal-bearing strata and the characteristics of the karst groundwater flow in the mining area, it concludes the geological structure model of the karst water system in the main coal mining area in north China into five types of monoclinic order type, monoclinic inversion type, parallel directional type, syncline basin type and fault block as well as other type. And it founds the foundation of establishing hydrogeological conceptual model for risk assessment of water inrush in coal seam floor.

Y. Zeng, *Research on Risk Evaluation Methods of Groundwater Bursting from Aquifers Underlying Coal Seams and Applications to Coalfields of North China*, Springer Theses, https://doi.org/10.1007/978-3-319-79029-9_6

[2] The main control factor system which affects the water inrush from coal seam floor is established. Combining with over years' a large amount of cases of water inrush in coal seam in China, it systematically analyzes the main factors influencing the water inrush in the coal floor from the influence of the water filling aquifer, floor of coal seam, underlying aquifer floor interval water rock section, geological structure, and excavation engineering disturbance on water inrush of floor. It respectively studies the water pressure and water abundance of water filling aquifer factors, and water resistance of ancient weathering crust; the equivalent thickness of effective aquiclude rock section and thickness of key layers in water resisting rock section factors; the distribution of fault and fold, distribution of collapse column, distribution of end points and intersection points of fault and fold, scale index of fault; the thickness of brittle rock of mine pressure developing zone, raising developing zone and damage zone in disturbance factors of mining engineering. Aiming at the above main factors, a set of main control factors index collection, factor index quantification and immeasurable steel processing methods are improved. The main control system of floor water inrush is established, which lays the basis for multi—factor information fusion model.

[3] The new method of water inrush evaluation of coal seam floor is studied, that is, the vulnerability index method based on partition weighting theory. It firstly uses the information fusion technology to study the multi-source information fusion architecture and fusion algorithm of water inrush evaluation, and establishes the JML—based water inrush evaluation information fusion model based on GIS, and utilizes AHP and ANN technology to study the method of determining the weight of each main control factor. In view of the above two researches, this paper summarizes the vulnerability evaluation method based on the theory of constant weight. Built on the method of vulnerability evaluation that based on constant weight theory, the K-mean value clustering method in dynamic clustering is used to classify the variable range thresholds of the main control factors, and the constructs the state variable weight vector and the determined the theoretical approach of adjustment parameters. A new method for risk assessment of water inrush from coal floor is studied, which is based on vulnerability index method. This method can not only consider a variety of main control factors and their corresponding different weights, but also can quantitatively determine the corresponding weight of the same factor in different state values, which solves the key technical problems in vulnerability evaluation and evaluation in coal seam floor water inrush.

[4] Research on engineering application of new method for risk assessment of floor water inrush. The vulnerability index method based on partitioned variable weight theory is applied. The vulnerability evaluation of the Ordovician limestone water inrush in the #9 coal seam of the monoclinic order type Xiandewang Coal Mine is carried out; the risk assessment of the Ordovician limestone water inrush in the #10 coal seam of Wangjialing Coal Mine is carried out. And compared with the traditional water inrush coefficient method and the evaluation result based on the constant weight vulnerability index method. The results

show that the water inrush coefficient method only considers the water pressure of the aquifer and the thickness of the aquifer, and the evaluation result is only dangerous and safe, the evaluation accuracy is low; while the traditional vulnerability index method in the evaluation process, the index with the same water inrush main control factors has a certain value for weight, but the method does not change with the change of water main control factors, and cannot match the actual situation.

6.2 Additional Research and Outlook

[1] To construct the main control factor partition state vector, the first requirement is to determine the main control factor variable range (threshold). However, the determination of variable threshold is now a difficult point. There is no unified analyzed determined method, although this paper, according to the characteristics of clustering distribution in the spatial distribution of the study area of different state value of the same factor, proposes a method to determine the threshold of variable range by using K-mean value clustering.

[2] The key step in constructing the variable weight vulnerability evaluation model of water inrush in coal seam is to determine the adjustment parameters of the model. These parameters can control and adjust the variable weight effect of the weight, and play a corresponding "punishment" "incentive" role. However, due to the study of variable weight theory is not perfect, there are few studies on the determination of model parameters. Although this paper puts forward a method to determine the adjustment parameters, it still rely on a comprehensive consideration of various factor index value or consulting the relevant experts to determine in selecting the ideal weighting weight value of evaluation of unit. The key point in the future is to determine a set of quantitative methods that can realize the adjustment parameters of the expected adjustment effect.